LIQUID SEMICONDUCTORS

LIQUID SEMICONDUCTORS

MELVIN CUTLER

Department of Physics
Oregon State University
Corvallis, Oregon

ACADEMIC PRESS　　　New York　San Francisco　London　1977
A Subsidiary of Harcourt Brace Jovanovich, Publishers

6 232-3945 √

ACADEMIC PRESS, INC.
111 Fifth Avenue, New York, New York 10003

United Kingdom Edition published by
ACADEMIC PRESS, INC. (LONDON) LTD.
24/28 Oval Road, London NW1

Library of Congress Cataloging in Publication Data

Cutler, Melvin,   Date
    Liquid semiconductors.

    Bibliography: p.
    1.   Liquid semiconductors.   I.   Title.
QC611.8.L5C87   537.6'22   76-27436
ISBN 0–12–196650–X

PRINTED IN THE UNITED STATES OF AMERICA

CONTENTS

Contents

Appendix C   Bond Equilibrium Theory   208

References   213

PREFACE

The definition of liquid semiconductors as a research area seems to date back to a review article by Ioffe and Regel (1960). Information and understanding about liquid semiconductors has since advanced to the point where various investigators are beginning to propose or consider some general models for liquid semiconductor behavior. A review of this subject thus seems worthwhile in order to consolidate critically the existing information and to provide a stronger basis for further progress.

The main goal of this work is to present a discussion of the present status of the field of liquid semiconductors that will be of value for present and future workers in this and related areas. The book also provides an opportunity to set forth my own views about promising approaches for understanding liquid semiconductor behavior and, I hope, to stimulate much-needed investigations of these materials by scientists who have expertise in a number of methods not yet extensively applied to liquid semiconductors, such as NMR or diffraction techniques. There are many fertile opportunities for valuable and interesting work that have been obscured by the lack of a comprehensive source of information, and perhaps this monograph may correct the situation. The properties and nature of liquid semiconductors suggest many concepts and problems that are of value and interest to scholars interested in the physics or chemistry of condensed matter, and in this volume these concepts and problems are made available to readers with only general interest in or curiosity about the subject.

In keeping with these purposes, the material is presented at a level accessible to a reader moderately knowledgeable in solid state physics. This requires a compromise between testing the patience of better-informed readers and asking others to refer back to more basic sources. No effort is made to be encyclopedic in references to experimental and theoretical work on the subject, as might suit a reader with some detailed interest in one specific topic. The references should be sufficient to provide good starting points for an investigation of the literature. Inevitably, the extent of discussion will be colored by my own background and interests in a way that will be obvious to a perceptive reader.

The central area of interest is the electronic behavior of liquid semiconductors. This behavior is the most striking and unique property of these materials and a key to understanding them. This is not to belittle structural, thermochemical, and other aspects of the subject, for which unfortunately there is a great deficiency of information. Thus, it is not possible to include as much material about nonelectrical aspects. At this point, it is appropriate to refer to the monograph by Glazov, Chizhevskaya, and Glagoleva (1969), "Liquid Semiconductors," which treats the subject from a physicochemical point of view. In the present book, chemical and metallurgical aspects are discussed in a way that provides a rounded presentation, but the emphasis is on electronic aspects.

Liquid and amorphous semiconductors may be regarded as different facets of a broader domain of study, that of disordered electronically conducting systems. This common domain has become an increasingly active field of research, and the rate of progress in it has recently accelerated. A major part of this new activity is concerned with solids, but the theoretical developments have broad application, and there are other overlapping areas. The reader is referred to recent monographs on the broader subject of disordered systems: "Electronic Processes in Non-Crystalline Materials," by Mott and Davis (1971), and "Amorphous and Liquid Semiconductors," edited by J. Tauc (1974). An earlier treatise that has contributed to the subject is A. I. Gubanov (1965), "Quantum Electron Theory of Amorphous Conductors." Another important source of information is to be found in proceedings of international conferences, which have been held every two years on amorphous and liquid semiconductors, the most recent one (the sixth) at Leningrad in 1975.

The organization of this monograph is determined by the state of development of the field and my own approach (as an experimentalist) toward understanding the subject. The field is in a rather early stage of development, in which the appropriate concepts for proper understanding are only beginning to be apparent. Consequently the most concrete facts are the experimental data and the questions they raise. Accordingly, the chapters fall into three main sections following the introduction, which serves to define the field of investigation and consider its relationship to other classes of materials. The first section (Chapters 2–4) is a systematic review of experimental information, with the purpose of uncovering some of the basic questions posed. Chapters 2 and 3 concentrate, respectively, on physical and on chemical or metallurgical information. It also seems desirable to include a chapter (Chapter 4) that reviews experimental methods and problems having particular significance for the study of liquid semiconductors. The second section (Chapters 5 and 6) discusses the theoretical basis for interpreting the experimental behavior. Chapter 5 considers the theoretical and conceptual basis for understanding or describing the electronic structure, and Chapter 6 the theory for interpreting the various experimental measurements. In this section, we emphasize those results that seem to be fairly well established, and we defer discussion of more tentative or speculative theories. The final section (Chapters 7 and 8) aims at examining existing information about various liquid semiconductors in the light of existing theories and concepts in order to determine what specific conclusions can be derived. The extent to which this can be done depends strongly on the amount and diversity of reliable data available for each liquid semiconductor system, as well as how the results have lent themselves to coherent explanation. Thallium–tellurium alloys are well favored by both these criteria, and Chapter 7 is devoted to this system and pure tellurium. Many other alloy systems have elements of behavior that can be profitably compared with or contrasted to that of Tl–Te. Discussions of other alloy systems are conveniently grouped together in Chapter 8.

ACKNOWLEDGMENTS

It is appropriate to acknowledge and express my appreciation for the stimulus to this work provided by interactions with others in this field. The provocative data obtained by former associates and students in my laboratory—Charles E. Mallon, James F. Leavy, Marshall B. Field, James M. Donally, Michael M. Dahl, and Robert L. Petersen—have provided reliable and fertile material for use in trying to formulate explanations for liquid semiconductor behavior. The development of my understanding of the subject has benefited from discussions—sometimes lively ones—with many other investigators in the field. I am indebted to J. C. Perron and D. N. Lee for making available to me unpublished work from their doctoral theses. I am grateful to William W. Warren, Jr., Jan Tauc, and John A. Gardner for critical comments on parts of the manuscript. This work owes its greatest debt to Professor Sir Nevill Mott. Besides his many important direct contributions to the subject, which will be evident in what follows, this work has benefited from my many interactions with him, including a sabbatical year in his laboratory, which have encouraged and stimulated my study of the subject and illuminated my understanding.

CHAPTER 1

INTRODUCTION

Semiconducting liquids are a poorly understood class of materials as compared to others, such as liquid metals or molten salts, where first-approximation models exist (for example, the free electron gas model for metals, or the Born model for ionic materials). There is no simple model which is generally accepted as a first approach for explaining the behavior of liquid semiconductors, and the development of the appropriate concepts for doing this poses a challenging problem.

The physical character of liquid semiconductors puts them in an intermediate position between several other classes of materials: liquid metals, molten salts, molecular liquids, and amorphous solids. It is often difficult to define the boundaries precisely. Some clarification of the interrelations will be one of our goals, and it will be pursued to some extent in this introductory chapter. Through these relationships, present understanding of other materials can be used in solving problems concerning liquid semiconductors. Conversely, an improved understanding of liquid semiconductors can be expected to add a deeper insight into the character of the better known substances.

To place the discussion on a more solid footing, it seems best to start by defining liquid semiconductors. The definition used by Ioffe and Regel in their pioneering review article (Ioffe and Regel, 1960) serves as a rough starting point: Liquid semiconductors are electronically conducting liquids with electrical conductivities less than the usual range for liquid metals. This distinction from molten salts and molecular liquids, which do not conduct electronically, seems simple, but it will require some elaboration. At the other end of the conductivity range, the line of distinction from liquid metals is arbitrary and has been a subject of some controversy. Ioffe and Regel have suggested that liquids with electronic conductivities less than $\sim 10^4$ ohm^{-1} cm^{-1} (the value for liquid mercury) should be regarded as liquid semiconductors. We discuss this in more detail below.

Before discussing further the boundaries of liquid semiconductors with other classes of liquids, let us outline briefly what kinds of materials are commonly regarded as liquid semiconductors. Among the elements, molten selenium and molten tellurium fall into this category. Other elements which are semiconductors in the crystalline state, such as Ge and Si, become metals on fusion. The same is true for many compound semiconductors such as the III–V compounds. This change from a semiconductor to a metal on melting has been correlated by Ioffe and Regel (1960) with a decrease in the atomic volume. Thus the well-known association of crystalline semiconducting behavior with a large atomic volume and a small coordination number seems to be preserved in the liquid state. Many other compounds, such as In_2Te_3, increase in volume on melting or else have relatively small decreases in volume, but nonetheless have electrical properties associated with semiconducting liquids. This relationship is discussed in more detail in Section 3.2.

Early studies of liquid semiconductors were based mainly on observing what happens to the electrical behavior when crystalline semiconductors are melted, but since liquids can have wide ranges of stoichiometry, liquid semiconductors are better regarded as alloy systems in which the composition is a continuous variable. In this frame of reference, liquid semiconductor systems include Se–Te, Tl–Te, Tl–Se, Ag–Te, Mg–Bi, V–O, and many other binary systems. Obviously, very many tertiary or more complicated systems would fall in this category over some ranges of composition, but at the present stage of development, the added complexity of more than one composition variable does not often help to shed light on the basic questions. Consequently, there have not been very many studies of tertiary systems. Tertiary systems which have been studied profitably include ones which can be regarded as pseudobinary systems such as Ga_2Te_3–Ga_2Se_3, and systems in which a third element is added as a "doping" agent in a binary system of fixed composition such as Tl–Te–M and Tl–Se–M, where M represents a relatively small amount of an element such as Cd, In, or Sb. The reader may have noted the frequent occurrence of a chalcogenide element in systems which have been mentioned. Most known liquid semiconductor systems do contain a group VIB element. Although this correlation is not universal, it is a significant characteristic of liquid semiconductors which needs to be understood.

The definition of the boundary between liquid semiconductors and insulating liquids such as molten salts and molecular liquids, although simple at first sight, shows some complexity on deeper examination. It is clear that electronic conduction and ionic conduction can occur in parallel in molten salts, and one or the other may dominate in some systems depending on the temperature or other conditions. As noted by Ioffe and Regel (Ioffe and

Regel, 1960), the electrical conductivity σ of an ionic liquid is unlikely to be larger than ~ 1 ohm^{-1} cm^{-1} because of the limited ionic mobility, so that liquids with higher conductivities must conduct electronically. When σ is lower than that, other considerations, such as measured ionic transport numbers or the chemical nature, must be used to establish the extent of electronic conductivity.

A substance such as liquid selenium has a strongly temperature-dependent electrical conductivity which may be as low as 10^{-7} ohm^{-1} cm^{-1} near the freezing point. Its chemical nature indicates that it is a covalent rather than an ionic substance. A possible mechanism for electronic transport in such a material is by hopping of electrons in localized states which may be charged. If so, the charged molecular complex can drift in an electric field, thus giving it the character of an ionic liquid, albeit a dilute one. Furthermore, recently developed concepts about the nature of electronic motion in localized states suggest the possibility of a wide range of correlation between the motion of the charge and the motion of the associated molecular complex. In view of this, the distinction between ionic transport and hopping transport of electrons in localized states may not be sharp in liquids. This is one of the subjects which needs to be clarified in future studies.

There are some types of electronically conducting liquids other than liquid metals which are not conventionally regarded as liquid semiconductors. One of them consists of solutions of alkali metals or alkaline-earth metals in ammonia or related liquids (Cohen and Thompson, 1968). These are low temperature molecular liquids in which electrons can occur in localized or extended states, depending on the concentration of the metal ions. A second type includes a number of molten halide salts which dissolve large amounts of the metal, such as La–LaCl$_3$, Bi–BiI$_3$, and molten alkali halides with excess metal (Bredig, 1964). The distinction between metal solutions in molten salts and liquid semiconductors is mainly conventional. All graduations can be found in the ionicity between these substances and systems conventionally regarded as liquid semiconductors, so that only quantitative distinctions can be made. It will be seen later that a "dilute metal" model provides a useful way to explain the behavior of many liquid semiconductor systems in certain ranges of composition, and this is not qualitatively different from reasonable models for these two types of systems. Therefore there is probably considerable overlap.

The boundary with liquid metals has been the subject of some controversy, and it represents a topic to be clarified through study of the nature of the intermediary materials rather than be debated. Semiconducting liquids in the high conductivity range ($\gtrsim 100$ ohm^{-1} cm^{-1}) are statistically degenerate and obey Fermi–Dirac statistics. In this sense, they are metals, and the name liquid semiconductor is partly a historic accident. Probably one reason they were called semiconductors is that in many cases they show a strong increase in conductivity with temperature, in contrast to typical liquid metals, and like solid semiconductors in the classical (now obsolete) definition. Recent studies (Cutler, 1971a) indicate that in some cases this sensitivity to temperature cannot be ascribed to excitation of carriers across a gap or out of traps, but must reflect a change in chemical structure with temperature. Therefore the mechanism for the temperature sensitivity of the electrical conductivity may be different from conventional semiconductors. Ioffe and Regel (1960) emphasized the fact that the mean-free path in semiconducting liquids is of the order of the interatomic distance, in contrast to liquid metals. It has become established that the Ziman theory for electronic transport, based on weak scattering, provides a good description of transport in liquids with electrical conductivity $\sigma \gtrsim 10^4$ ohm^{-1} cm^{-1}. On the other hand, other approaches to transport theory, based on strong scattering, seem to be applicable for $\sigma \gtrsim 2500$ ohm^{-1} cm^{-1}. (This subject is discussed more thoroughly in Section 6.1.) Therefore liquids with $\sigma \gtrsim 10^4$ ohm^{-1} cm^{-1} apparently differ in some important respects from the usual liquid metals, and as the conductivity is decreased, some of the classical concepts of semiconducting behavior seem to become increasingly pertinent. Certainly, in the range where $10,000 \gtrsim \sigma \gtrsim 1000$ ohm^{-1} cm^{-1}, investigators whose main concern is liquid metals or liquid semiconductors may feel equally at home.

The nature of the boundaries between liquid semiconductors and ionic, molecular, or metallic liquids has many similarities to solids. However, the pecularities of the liquid state lead to important differences. As noted already, diffusive motion of atoms in the liquid may play a special role in electronic transport when electrons are in localized states. Another difference, which arises from the wide range of stoichiometry which can occur in the liquid state, is that the electronic structure changes continuously in response to changes in chemical composition. We believe this to be the most important attribute of liquid semiconductor behavior. This characteristic offers an opportunity for developing a deeper understanding of one of the basic problems in the physics and chemistry of condensed matter, that is, the mutual relation between the electronic structure and the atomic or chemical structure of matter. It seems likely that the chemical structure of many liquid semiconductor systems is based on covalent bonding, but in contrast to the common molecular liquids, the high temperature and chemical environment is one in which the resulting molecular species are not well-defined (especially at present). Thus a rapidly changing dynamic equilibrium between various atomic arrangements probably plays a role in determining the effect of changes in temperature or composition. In addition to this, the binding which occurs in liquid semiconductor systems can range continuously to the extremes of ionic or metallic binding. Thus we are concerned with a type of matter which is an unusual "soup" from a conceptual as well as physical point of view, for which many of the usual concepts or models are not adequate. The resulting problems seem to call for attention by theoretical chemists as well as physicists, and the resolution seems likely to provide valuable insights for both fields.

Liquid semiconductors also have a special relationship to solid amorphous semiconductors. The fact that there is a liquid rather than a solid has no significance in many aspects of electronic behavior, since the time scale for many types of electronic motion is much faster than for atomic motion, and the same concepts or theories are applicable. The aspects of behavior in which the fluidity is significant are mainly those already mentioned, and they are not the dominant element in electrical behavior. However, there are many practical or quantitative differences which cause a considerable difference in emphasis. For one thing, most of the systems are liquid in the high temperature range of 400–1000°C, and no semiconducting melts are known below 170°C. Thus, low temperature types of studies are out of the picture. This, together with certain chemical factors to be discussed later, leads to the result that liquid semiconductors typically have much higher concentrations of electronic carriers than amorphous solids. As indicated earlier, many liquid semiconductors are nearly (or completely, depending on your point of view) metallic, whereas most typical amorphous semiconductors are more closely related to insulators. In terms of the electronic structure, this means that amorphous solids very frequently represent situations where the Fermi energy is well separated (in units of kT) from energies where extended states occur (conducting bands), while liquid semiconductors typically have the Fermi energy near or within conducting bands. This is not by any means a universal dichotomy; liquid selenium is a very respectable insulator, and one could also lay claim to liquid sulfur, an excellent insulator, as a liquid semiconductor. Conversely, many amorphous solids have been made with metallic or near-metallic properties.

From a practical point of view, some other distinctions between amorphous solids and liquid semiconductors should be mentioned. Assured achievement of homogeneity is usually a relatively simple experimental problem for liquids, whereas preparation of reproducible amorphous solid samples is difficult except possibly for natural glasses, which occur only in restricted ranges of compositions. On the other hand, as already noted, the fact that liquids must be studied at high temperatures presents special experimental problems and places important limitations on what can be done experimentally.

CHAPTER 2

EXPERIMENTAL INFORMATION ON PHYSICAL PROPERTIES

In this chapter and Chapter 3, we review existing experimental information about liquid semiconductors. The present chapter is concerned with physical properties, and the following one with chemical and physiochemical properties. Our purpose is to set forth what is generally known experimentally about these substances, and the information presented will often be representative rather than exhaustive. In later chapters, we consider the applicability of present theoretical knowledge to existing experimental information.

Of all the physical properties, the electronic transport parameters are the most distinctive ones for liquid semiconductors. Therefore, it is not surprising that a large fraction of existing information consists of studies of the electrical conductivity σ, which is the most easily measured transport parameter. Measurements of the thermopower S (Seebeck coefficient), Hall coefficient R_H, and thermal conductivity κ are carried out with rapidly decreasing frequency. We devote a section to each of these measurements as well as one on the magnetic susceptibility χ_M. Studies of nuclear magnetic resonance and optical properties are so sparse that it will be more appropriate to combine the presentation of existing information with discussion of their interpretation. This will be done in Chapters 6–8.

2.1 ELECTRICAL CONDUCTIVITY

Many of the early studies of liquid semiconductors consisted simply of measurements of the electrical conductivity σ of molten versions of solid semiconductors (Ioffe and Regel, 1960). The initial concern was to determine which solids remain semiconductors in the liquid state. The reported information is frequently limited to a curve for σ as a function of temperature T, taken through the melting point. Figure 2.1 shows some typical curves of this sort. As mentioned earlier, many solid semiconductors, such as

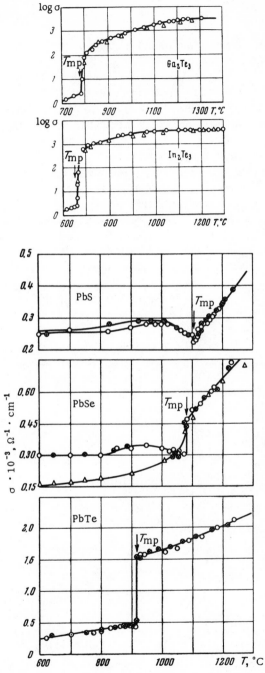

Fig. 2.1. Behavior of σ on melting crystalline semiconducting compounds (Glazov *et al.*, 1969).

germanium or indium antimonide, become metallic in the liquid state, as indicated by large values of σ ($\gtrsim 10^4$ ohm^{-1} cm^{-1}). Among the melts which are less metallic, a very wide range of electrical conductivities is found. The slope of $\sigma(T)$ is frequently large and positive, and this is more likely to be true when σ is small. This is roughly similar to what is observed in crystalline semiconductors. In Table 2.1, we list the electrical conductivities of a number of molten semiconductors which are representative. A more exhaustive list has been published by Allgaier (1969).

The phase diagrams of binary alloy systems corresponding to semiconducting melts frequently show compounds at more than one composition. This in itself suggests that there may not be a unique stoichiometry for the liquid phase, and if there is one, a relationship to that of solid compounds

TABLE 2.1

Properties of Molten Semiconductors[a]

(1) Substance	(2) Relative change in density on melting (%)	(3) Electrical conductivity of liquid (ohm^{-1} cm^{-1})	(4) Thermopower of liquid (μV/deg)
Si	10	12,000	—
Ge	4.15	14,000	0
AlSb	12.9	9,900	−60
GaSb	8.2	10,600	0
InSb	12.5	10,000	−20
In$_2$Te$_3$	−3.98	60[b]	30
Ga$_2$Te$_3$	−4.98	50[b]	45[c]
CuI	−9.0	3.5	490
GeTe	−6.7	2,600	21
SnTe	−4.9	1,800	28
PbTe	−3.1	1,520	−10
PbSe	−6.2	450	−60
PbS	−8.8	220	−200
Sb$_2$Te$_3$	−3.2	1,850	11
Bi$_2$Te$_3$	−3.2	2,580	−3
Bi$_2$Se$_3$	−4.1	900	−35
Se	−15.2	$\sim 10^{-6}$	1200[d]
Te	−5.3	2,200[e]	45[e]

[a] Unless otherwise noted, the data are from Glazov *et al.* (1969).
[b] Lee (1971).
[c] Valient and Faber (1974).
[d] Gobrecht *et al.* (1971b).
[e] Perron (1967).

is not essential. In any case, studies of properties of molten alloys A_xB_{1-x} generally show a continuous variation in behavior as a function of the composition parameter x in addition to T. This second experimental dimension is very important in studying liquid semiconductors.

A number of binary systems have been studied as a function of composition as well as temperature. The system Tl_xTe_{1-x} has been studied by Cutler and coworkers (Cutler and Mallon, 1965, 1966; Cutler and Field, 1968; Cutler and Petersen, 1970), Enderby and Simmons (1969), Nakamura and Shimoji (1969), and by Kazandzhan and coworkers (1971, 1972). The very good agreement between the results from these different laboratories assures their reliability. Figure 2.2 shows the dependence of σ on temperature for a number of compositions. Two distinctive types of behavior are evident. For $x > \frac{2}{3}$, $d\sigma/dT$ is small and positive, and σ changes rapidly with composition. For $x < \frac{2}{3}$, $d\sigma/dT$ is large and σ changes less rapidly with composition. The nearly constant slope of $\ln \sigma$ when plotted versus T^{-1} indicates a thermally activated process. Other significant aspects of the behavior are revealed by the resistivity isotherms $\rho(x)$, which are plotted in Fig. 2.3a. The maximum at $x = \frac{2}{3}$ is reminiscent of nonstoichiometric doping of a compound semiconductor, in this case, Tl_2Te. The phase diagram for the alloy system

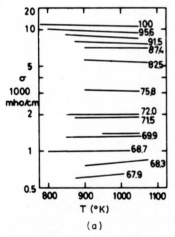

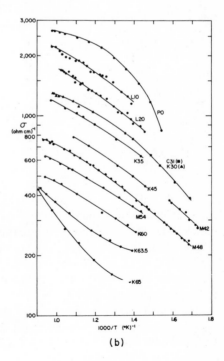

Fig. 2.2. $\sigma(T)$ for various compositions of Tl–Te alloys. The numbers indicate the compositions in at% Tl (a) Tl–rich alloys (Cutler and Petersen, 1970), (b) Te–rich alloys.

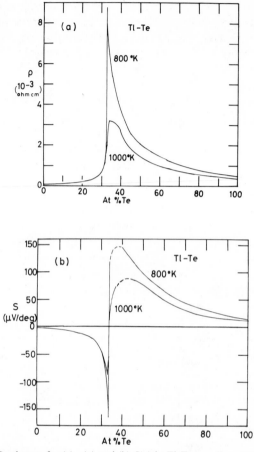

Fig. 2.3.   Isotherms for (a) $\rho(x)$ and (b) $S(x)$ in Tl_xTe_{1-x} alloys (Cutler, 1971a).

Tl–Te, shown in Fig. 2.4, has no solid phase Tl_2Te, and there are several other compositions where compounds do occur: TlTe, Tl_2Te_3, and a non-stoichiometric phase with an approximate composition Tl_5Te_3. The asymmetry of the isotherms about the stoichiometric composition in Fig. 2.3a indicates that excess Tl increases the conductivity much more effectively than excess Te. This feature is significant in determining possible explanations for the electrical behavior.

We find it useful to characterize the two modes of electrical behavior, which are frequently observed in other alloys as well as Tl–Te, as M-type or S-type. By M-type (alluding to metals) we mean that the electrical prop-

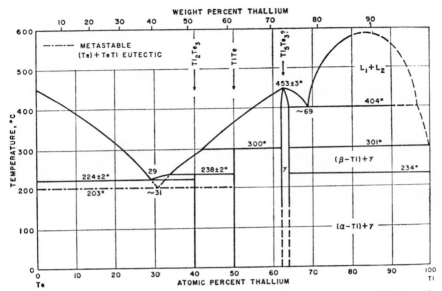

Fig. 2.4. Phase diagram for the binary system Tl–Te (Hansen and Elliot, see Hansen and Anderkov, 1958).

erties are weakly dependent on temperature, and in S-type behavior (alluding to semiconductors) the electrical properties depend strongly on temperature. A number of authors have categorized liquids by other criteria which should be noted here. Enderby and Collings (1970) have classified liquids as type I or type II, according to whether their electrical properties show irregular changes with composition (II) or not (I); they associate the latter with numerical values of the transport parameters with the ranges characteristic of typical metals. Category II corresponds roughly with what we call liquid semiconductor behavior, but the boundary which we suggest ($\sigma \gtrsim 3000$ ohm^{-1} cm^{-1}) is much higher. Allgaier (1969, 1970) has set up ranges A, B, and C with boundaries in terms of σ: $\sigma \sim 5000$ ohm^{-1} cm^{-1} separates A and B and $\sigma \sim 100$ ohm^{-1} cm^{-1} separates B and C. He has listed a large number of liquids and has made correlations with other properties according to their range. There has been a tendency to use these classifications to excessively characterize the nature of the liquid as opposed to its behavior. Although there is a correlation between the two, it seems to be an unfortunate emphasis, since the properties of many alloys vary continuously between different ranges of behavior with changes in composition or temperature. Allgaier's classification coincides roughly with a characterization of electrical behavior introduced by Mott (1971), with boundaries at $\sigma \sim 3000$ and

~ 300 ohm^{-1} cm^{-1}, on the basis of theory for transport and electronic structure. We discuss this theory in more detail in Chapters 5 and 6, and we shall follow Mott's criteria for characterizing transport behavior.

Returning to the discussion of alloy systems, we consider next the information about the alloy Se$_x$Te$_{1-x}$. The electrical conductivity has been studied by a number of authors, most comprehensively by Perron (Perron, 1967; Cutler and Mallon, 1962; Mahdjuri, 1975; Ioffe and Regel, 1960, and references therein). Perron's $\sigma(T)$ curves for various compositions are shown in Fig. 2.5. It is seen that σ changes monotonically with composition from near-metallic values for tellurium to extremely small values at large x. A second noteworthy characteristic is that σ has a large temperature coefficient reflecting an activation energy over wide ranges of x and T. The range of this S-type behavior is correlated most closely with the magnitude of σ. It occurs when σ is less than ~ 2000 ohm^{-1} cm^{-1}, and $d \ln \sigma / dT^{-1}$ tends to be constant

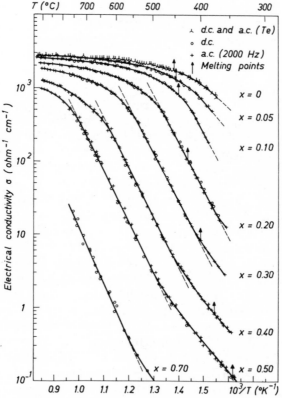

Fig. 2.5. $\sigma(T)$ curves for various compositions of Se$_x$Te$_{1-x}$ alloys (Perron, 1967).

TABLE 2.2

Semiconductor Behavior of Molten Binary Systems

(1)	(2)	(3)	(4)	(5)	(6)	(7)	(8)
		Minimum value		Inversion			
	Range	σ_{min} at x_{min}		of sign of	M-type	S-type	
M_xX_{1-x}	of x	(σ in ohm^{-1} cm^{-1})		S at x	range	range	References[a]
		σ	x				
Cu–Te	0–0.7	500	0.67	None	>0.5	<0.5	1
Ag–Te	0–0.7	100	0.67	0.67	>0.33	<0.33	1
Sn–Te	0–0.77	1400	0.5	0.50	>0.5	<0.5	1, 2
Pb–Te	0–1	1100	0.5	0.5	Most of range	~0.5	2, 3
Ge–Te	0–1	2300	0.3	>0.8	0–1	—	2, 3
Zn–Te	0.3–0.7	50	0.50	—	—	—	2
Cd–Te	0.3–0.7	50	0.50	—	—	—	2
Ga–Te	0–1	13	0.40	0.7–0.8	>0.4	<0.4	2, 3
In–Te	0–1	60	0.40	~0.65	>0.67	<0.67	2, 4–6
Sb–Te	0–1	1600	0.4	None	>0.67	<0.67	2, 6
Bi–Se	0.3–0.8	—	—	—	>0.45	<0.45	2
Bi–Te	0–1	1200	0.2	~0.9	>0.4	<0.4	2, 6
Mg–Bi	0–0.8	45	0.6	0.6	Most of range	~0.6	7–9
Pb–Se	0.4–0.75	—	—	—	—	—	9
Sn–Se	0.25–0.54	225	~0.5	—	—	<0.4	9
In–S	0.4–0.57	—	—	—	—	≳0.5	10
Tl–Te	0–1	70	0.67	0.67	>0.67	<0.67	11–18
Tl–Se	0–1	5	0.67	0.67	>0.67	<0.67	16, 19, 20
Tl–S	0.45–0.72	3	0.67	0.67	>0.67	<0.67	21, 22
Te–Se	0–1	—	—	—	—	0–1	23–26, 32, 33
Cs–Au	0.43–1	3	0.50	—	Most of range	~0.5	27, 28
Li–Bi	0.73–1	500	0.25	—	Most of range	~0.75	29
As–Se	0.3–0.5	—	—	—	—	0.3–0.5	30
As–Te	0–0.55	—	—	—	—	0–0.55	31

[a] *References*: (1) Dancy (1965); (2) Glazov *et al.* (1969); (3) Valient and Faber (1974); (4) Nino-miya *et al.* (1973); (5) Popp *et al.* (1974); (6) Blakeway (1969); (7) Enderby and Collings (1970); (8) Ilschner and Wagner (1958); (9) Glazov and Situlina (1969); (10) Morris (1971); (11) Cutler and Mallon (1965); (12) Cutler and Mallon (1966); (13) Cutler and Field (1968); (14) Cutler and Petersen (1970); (15) Enderby and Simmons (1969); (16) Nakamura and Shimoji (1969); (17) Kazandzhan *et al.* (1971); (18) Kazandzhan *et al.* (1972); (19) Regel *et al.* (1970); (20) Petit and Camp (1975); (21) Nakamura *et al.* (1974); (22) Kazandzhan and Tsurikov (1974); (23) Perron (1967); (24) Cutler and Mallon (1962); (25) Ioffe and Regel (1960); (26) Mahdjuri (1973); (27) Hoshino *et al.* (1975); (28) Schmutzler *et al.* (1976a, b); (29) Steinleitner *et al.* (1975); (30) Hurst and Davis (1974); (31) Oberafo (1975); (32) Mahdjuri (1975); (33) Andreev (1973).

when $\sigma \gtrsim 300$ ohm^{-1} cm^{-1}. It will be seen that many binary systems consisting of either Te or Se plus a more electropositive element have a singular dependence of σ on composition, as in Tl–Te. The chemical similarity of Se and Te suggests that these two elements play a similar role in these binary systems, yet their electrical properties in the chalcogenide-rich range of composition are widely different. This difference is exemplified by the behavior of the Se–Te alloys.

Information about other binary systems is generally much less extensive, and in some cases the accuracy is dubious. We summarize the information in Table 2.2 together with references. Among the more important systems are the In–Te and Ga–Te systems. Studies of In$_x$Te$_{1-x}$ as a function of x have been made by Blakeway (1969), by Glazov et al. (1969), and most recently by Ninomiya et al. (1973). We show isotherms $\sigma(x)$ from the work of Ninomiya et al. in Fig. 2.6. These isotherms resemble Tl–Te in having a minimum at a characteristic composition, but the minimum occurs at $x = 0.40$ rather than $x = \frac{2}{3}$, reflecting a valence 3 (In$_2$Te$_3$) rather than 1 (Tl$_2$Te). S-type behavior occurs, in this case, on both sides of the minimum σ. Solid compounds occur at a number of compositions including In$_2$Te$_3$ and In$_2$Te (Hansen and Anderkov, 1958). Glazov et al. (1969a) and Valient and Faber

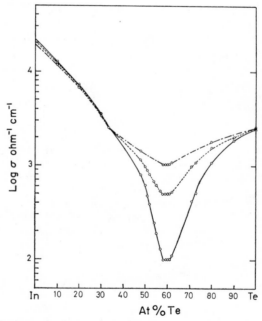

Fig. 2.6. σ isotherms for In–Te alloys: 700°C (———), 800°C (----), 900°C (—·—·—) (Ninomiya et al., 1973).

(1974) report measurements for Ga–Te, and the isotherms are very similar to In–Te.

Ag_xTe_{1-x} and Cu_xTe_{1-x} have been studied by Dancy (1965). The former has conductivity isotherms which are very similar to Tl–Te. The Cu–Te isotherms are somewhat different, and an S-type range occurs at $x < 0.5$ rather than $x < \frac{2}{3}$.

Studies of the electrical behavior of the alloy systems Tl–Se (Nakamura and Shimoji, 1969; Regel et al., 1970; Petit and Camp, 1975) and Tl–S (Nakamura et al., 1974; Kazandzhan and Tsurikov, 1974a) show that their behavior is very similar to Tl–Te. The main difference occurs in chalcogenide-rich compositions, where the electrical conductivities are much smaller. In the case of Tl_xS_{1-x}, σ drops to a low value at $x = \frac{2}{3}$, and remains constant at that value for smaller x.

The binary systems Sn–Te, Pb–Te, Ge–Te, Zn–Te, and Cd–Te all seem to have σ minima at $x = 0.5$. Where it is discernable in the reported data, M-type behavior seems to occur on the metal-rich side and S-type behavior on the chalcogenide-rich side of this composition. In each of these systems, the only solid binary compound occurs at the composition MTe. We show in Fig. 2.7 isotherms for Sn–Te from the work of Dancy (1965a).

Sb_xTe_{1-x} and Bi_xTe_{1-x} have been studied both by Glazov and coworkers (1969) and by Blakeway (1969). The σ isotherms in these two reports are only roughly similar to each other. The common features are M-type behavior on the metal-rich side, and S-type behavior on the other, but the boundaries are not clear. Also, the occurrence of minima in σ is not very certain. There are indications that it occurs at $x = 0.4$ for Sb_xTe_{1-x}, but the information on this point is ambiguous for Bi_xTe_{1-x}. Glazov et al. (1969) report $\sigma(T)$ curves for Bi–Se which indicate again some singularities in the middle range of compositions whose nature seems to be nebulous.

Some information on the systems Ni–S, Co–S, Cu–S, and Ag–S is provided by a study of Dancy and Derge (1963). In these alloys, conductivity minima occur in midranges of composition, but the results are not very precise.

Among binary alloy systems which do not include a chalcogenide element, it seems doubtful that any III–V alloys should be called liquid semiconductors. We do not list in Table 2.2 information reported for the Ga–Sb, In–Sb, or Al–Sb binary systems. All of them have $\sigma > 10^4$ ohm^{-1} cm^{-1}. Glazov et al. (1969) report minima for all three at $x = 0.5$, but results reported by Blakeway (1969) for In–Sb are quite different, and show no minimum. The alloy system Mg–Bi has long been known to have a composition Mg_3Bi_2 where σ has a strong minimum ($\ll 100$ ohm^{-1} cm^{-1}) indicating semiconductor behavior (Ilschner and Wager, 1958; Enderby and Collings, 1970). More recently, similar behavior has been suggested for Li–Bi with semiconductor behavior at composition Li_3Bi (Ioannides et al.,

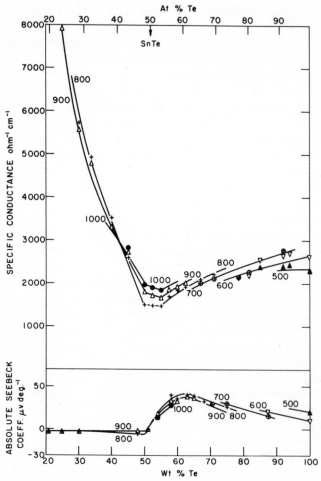

Fig. 2.7. Isotherms for σ and S in Sn–Te alloys: 1000°C (●), 900°C (△), 800°C (+), 700°C (⊕), 600°C (▽), 500°C (▲) (Dancy, 1965).

1973), and this has been since confirmed (Steinleitner *et al.*, 1975). More surprisingly, Cs–Au has also been found to have a strong minimum in σ (~ 3 ohm^{-1} cm^{-1}) near the composition CsAu (Hoshino *et al.*, 1975). All of these three systems show M-type behavior except in a narrow range of composition near the conductivity minimum.

In summary, a number of binary systems M_x–X_{1-x} have been studied, and among them a distinct minimum is frequently found in $\sigma(x)$. In most of them the electronegative element X is a chalcogenide, but there are several in which it is not. The minimum frequently occurs at a composition corresponding to a relatively stable solid compound, but in at least one (Tl–Te), this compound apparently does not occur as a solid. In most binary systems M-type regions can be discerned, and they tend to occur on the metal-rich side of the composition range. The quantity σ often changes rapidly with composition in the M-type range, as in Tl–Te. S-type regions are frequently observed in chalcogenide-rich binary alloys. Frequently, σ changes relatively slowly with composition in S-type ranges as in Tl–Te.

2.2 THERMOPOWER

The Seebeck coefficient (thermopower) S has been studied much less frequently than σ. This measurement is more difficult to make accurately, but it adds very significantly to information about a liquid semiconductor, particularly when the measurements of S are made at the same compositions as the measurements of σ.

Values of S for a number of molten semiconductor compounds are given in Table 2.1. Two features should be noted: (1) The magnitudes of S and σ are strongly correlated. When σ is in the metallic range ($\gtrsim 10{,}000$ ohm^{-1} cm^{-1}), S is generally small in magnitude ($\ll 100$ μV/deg). When σ is smaller, the magnitude of S systematically increases into the range $\gtrsim 100$ μV/deg; (2) Among liquids not in the metallic range of σ, S is frequently positive.

The theoretical relationship between σ and S will be discussed in detail in Section 6.1, but it is useful at present to note that the above-mentioned correlation between the magnitudes of S and σ is in accord with theoretical transport behavior when conduction is in a single band. If conduction occurs simultaneously in two bands, the magnitude of S may be small at the same time that σ is small. This is also observed in restricted ranges of composition of binary systems. (An alternative view, that the Fermi energy is near a minimum in the density of states is equally valid. We shall use the two-band terminology as a matter of convenience.)

As in the case of σ, studies of S as a function of composition in binary systems provide much more extensive information. The behavior of the thermopower of binary alloys is summarized in Table 2.2.

The $S(T)$ curves for $Tl_x Te_{1-x}$ at various values of x are shown in Fig. 2.8. It is seen that the temperature dependence is strong for $x < \frac{2}{3}$ and weak for $x > \frac{2}{3}$ as in the case of $\sigma(T)$; and that the magnitude of S has a single-band correlation with σ. As in the case of σ, the dependence on x is not monotonic. Figure 2.3b shows this dependence in the form of isotherms. An inversion in sign of S occurs at the composition $Tl_2 Te$, where σ is a minimum. This inversion can be understood in terms of a transition between transport in two different bands. Single-band transport occurs on either

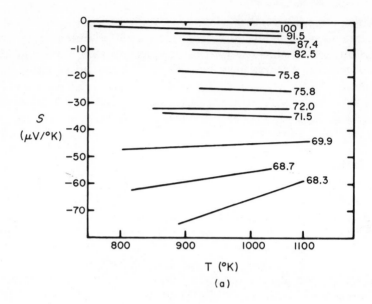

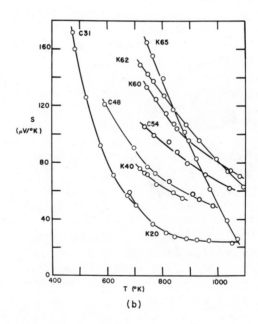

Fig. 2.8. $S(T)$ curves for various compositions of Tl–Te alloys. The compositions are given in at% Tl. (a) Tl–rich alloys (Cutler and Petersen, 1970), (b) Te–rich alloys.

side of the composition $x = \frac{2}{3}$, in which σ and S have their usual correlation. The region between the peaks in the magnitude of S is explained by ambipolar transport. In this range of composition, there is cancellation between the positive and negative contributions to S due to hole and electron transport.

The $S(T)$ curves for various values of x in $Te_{1-x}Se_x$ from the work of Perron (1965) are shown in Fig. 2.9. Comparison of these with the $\sigma(T)$ curves in Fig. 2.5 shows that the single-band correlation between the magnitudes σ and S carries over into the dependence on temperature as well as composition. This is frequently observed, and more specifically, it is often found, as in the present case, that when $\log \sigma$ depends linearly on T^{-1}, S does so also. The possible theoretical basis for this relationship will be discussed in Section 6.1. Aside from this, S has a monotonic dependence on x and T in Se_xTe_{1-x}, as does σ.

A correlation between a minimum in $\sigma(x)$ with an inversion in the sign of $S(x)$ similar to the one in Tl–Te is observed in the isotherms of a number of binary systems M_xA_{1-x}. As indicated in Table 2.2, this occurs in the systems Tl–Se, Tl–S, Ag–Te, Sn–Te, Pb–Te, and Mg–Bi. Figure 2.7 shows, as

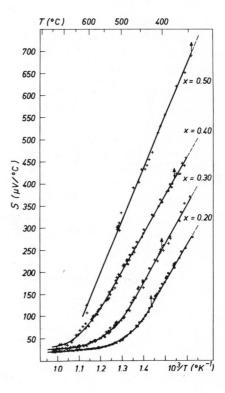

Fig. 2.9. $S(T)$ curves for several compositions of Se_xTe_{1-x} alloys; reference metals are silver ($\bigcirc$) and alumel ($+$); arrows indicate melting points (Perron, 1967).

an example, Dancy's (1965) isotherms of $S(x)$ and $\sigma(x)$ for Sn_xTe_{1-x}. In every case, S becomes positive in alloys rich in the electronegative component. In the alloys listed, there is good agreement between the compositions at which $\sigma(x)$ is a minimum and where $S(x)$ is zero.

The alloys In–Te, Ga–Te, and Ge–Te are distinctly different from those discussed so far in that the sign of S does not invert at the composition where ρ is maximum (Valient and Faber, 1974). Since S is positive at these compositions and negative for the pure metal, an inversion in sign must occur at larger values of x. The compositions at which an inversion in sign occurs are not exactly defined by the experimental data, and it apparently occurs at various values between $x = 1.0$ and 0.4, as shown in Table 2.2. In each case, the sign inversion occurs when σ is in the metallic range ($\sigma \gtrsim 2500$ ohm^{-1} cm^{-1}). The data for Bi–Te and Sb–Te seem to be less precise. These alloys seem to behave in a similar manner, with the exception that S is apparently positive for pure antimony and has no sign inversion (Blakeway, 1969). Studies of S in some transition metal sulfides do not seem to give clear indications about sign inversion (Bitler *et al.*, 1957; Dancy *et al.*, 1965).

In the discussion of σ, M-type and S-type ranges of behavior were noted in which the temperature dependence is weak or strong, respectively. In all the alloys in which the results are clear enough to decide on this point, the same sensitivity to temperature occurs in S as in σ, with the possible exception of compositions where S is changing sign. This is, of course, another manifestation of the single-band correlation between S and σ.

2.3 HALL EFFECT

Until the time period 1960–1965, published Hall measurements of electronically conducting liquids—mostly liquid metals—were of dubious reliability because of conflicting results. At that time, the concurrent work of several investigators succeeded in identifying and minimizing many of the important sources of experimental error (Cusack *et al.*, 1965). Subsequently, there has been a considerable improvement in the situation, and measurements of a number of liquid metals have been made with satisfactory agreement between different investigators.

An interesting consequence of the later studies of liquid metals was the observation that the theoretical relationship for the Hall coefficient R_H derived from the free electron model is accurately obeyed for a large number of pure liquid metals:

$$R_H = -1/ne, \tag{2.1}$$

where n is the density of valence electrons and e is the magnitude of the electronic charge (Faber, 1972). The fact that this relationship is obeyed better in liquids than in crystalline metals can be ascribed to the disorder in the liquid, which results in a smooth Fermi surface. This observation for pure liquid metals (but not alloys—see Faber, 1972) has colored the interpretation of subsequent investigations of liquid semiconductors.

The first reliable Hall measurements of molten tellurium date from the same period (Busch and Tieche, 1963). Contrary to an earlier report (Epstein et al., 1957), the sign of R_H is found to be negative at all temperatures, in contrast with the positive sign of the Seebeck coefficient. We show in Fig. 2.10 curves for R_H and the Hall mobility and μ_H versus T for molten tellurium. In conventional transport theory, the sign of both S and R_H is positive when transport is by holes, that is, empty states in a nearly filled band. Equation 2.1 is valid with a positive or negative sign for holes or electrons, respectively, over a wide range of values of n. When n is small enough for Maxwell–Boltzmann statistics to apply, an added factor b is introduced which is of the order of unity and whose value depends on the scattering mechanism (Beer, 1963). The discrepancy in sign between R_H and S might possibly be explained in terms of conduction in more than one band. Another possible interpretation, suggested by the accuracy of Eq. 2.1 for many liquid metals, is that

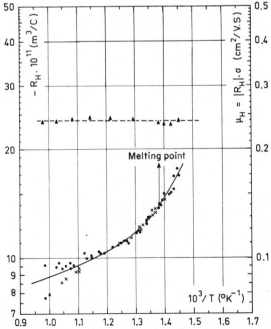

Fig. 2.10. Hall coefficient and Hall mobility of tellurium R_H at 115 Hz ($\bigcirc$) and 15 Hz ($\times$); $\blacktriangle$ indicate calculated points (μ_H) (Perron, 1970a).

this equation predicts correctly the sign and concentration of carriers for molten tellurium. In that case, a valence electron density is obtained which corresponds to about two electrons per tellurium atom instead of six, and this varies with temperature. This cannot be easily reconciled with a simple "metallic" model for the electronic structure of tellurium.

The question of the interpretation of the Hall coefficient of molten tellurium is part of a more general controversy concerning liquid semiconductors. Enderby and coworkers (Enderby and Walsh, 1965, 1966; Enderby and Simmons, 1969) have made Hall measurements in a number of semiconducting liquids Bi_2Te_3, Sb_2Te_3, SnTe, CuTe, AgTe), and they found negative values of R_H in all cases, even though these liquids have positive thermopowers. In many of them, Eq. 2.1 yields an electron density of approximately two per atom, and Enderby has suggested this as the proper interpretation. We shall consider the theoretical interpretation of the Hall coefficient in detail in Section 6.2, but note here that this question and the attendant controversy has played a significant role in the subsequent experimental and theoretical investigations.

Hall measurements have been made on a large number of semiconducting liquids, mostly at compositions corresponding to compounds. Allgaier (1969, 1970) has reviewed the data available at the time of his work, and he noted a tendency for the Hall mobility to decrease in a systematic fashion as the electrical conductivity decreases. This correlation is shown in Fig. 2.11. On going from typical metals ($\sigma \gtrsim 10^4$ ohm^{-1} cm^{-1}) to liquid semiconductors

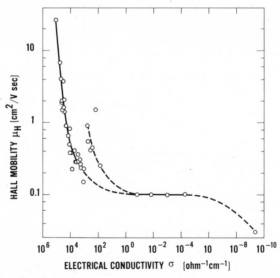

Fig. 2.11. Correlation between μ_H and σ in liquid conductors (Allgaier, 1970).

($\sigma \gtrsim 10^3$ ohm^{-1} cm^{-1}), the Hall mobility for various substances falls near a common curve which decreases with σ and levels off at the low end with mobilities of the order of 0.1 cm^2/V sec. It is risky to draw conclusions from this statistical type of treatment of experimental data. Since the sensitivity of a Hall measurement becomes very poor when $\mu_H \gtrsim 0.1$ cm^2/V sec, the apparent leveling off of μ_H near this value may reflect merely a selection of data, with materials with smaller Hall mobilities not being reported because of the experimental difficulty in detecting a Hall voltage.

A few binary alloy systems have been studied with the composition as a variable. Blakeway (1969) has reported measurements for the alloys Bi–Te and Sb–Te. Studies have also been made of Sb_2Se_3 in which appreciable amounts of several "doping" constituents were added, including Sb, Te, and In (Andreev and Mamadaliev, 1972; Regel et al., 1972). A study of Ge–Se alloys near the composition GeSe (Andreev and Mamadaliev, 1972) is of interest because it is the first (and perhaps only) example of a liquid semiconductor in which a positive Hall coefficient has been reported. R_H changes from negative to positive as Se is added, and the sign inversion occurs on the Se-rich side of GeSe. There is a similar sign inversion for the thermopower, but this occurs on the Ge–rich side of GeSe.

The Hall studies of the Tl_xTe_{1-x} system (Donally and Cutler, 1968, 1972; Enderby and Walsh, 1966; Enderby and Simmons, 1969) are particularly significant because of their extensive character together with other detailed experimental information on the system. The fact that measurements from two laboratories (by different experimental methods) are in good agreement also adds to their significance. It is found that μ_H ($= R_H\sigma$) is dependent of temperatures at all compositions except at $x = \frac{2}{3}$ (Donally and Cutler, 1972). It should be remembered that σ depends weakly on T for $x > \frac{2}{3}$, and it changes strongly with T for $x < \frac{2}{3}$ (see Fig. 2.2). In the latter range, R_H is proportional to σ^{-1} within the experimental error. We show in Fig. 2.12 a plot of μ_H versus x. There is a systematic change in μ_H with x, with a discontinuity at $x = \frac{2}{3}$. A more recent study of Tl_xSe_{1-x} (Kazandzhan and Selin, 1974) shows behavior very similar to Tl_xTe_{1-x}. These results are consistent with the two-band model for the electronic structure of Tl–Te that is suggested by the dependence of σ and S on x. The behavior at the composition Tl_2Te, where μ_H decreases with T, can be explained by a two-band formula:

$$\mu_H = \frac{\mu_n\sigma_n + \mu_p\sigma_p}{\sigma_n + \sigma_p}, \tag{2.2}$$

where μ_n and μ_p (n and p refer to electrons and holes, respectively) are both independent of T with values 0.1 and 0.4 cm^2/V sec, respectively, so that μ_H lies within this range. σ_n and σ_p depend on T, and the ratio σ_n/σ_p increases

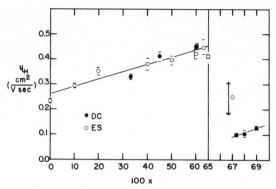

Fig. 2.12. Dependence of Hall mobility on composition in Tl_xTe_{1-x} alloys. The double arrow indicates the range of μ_H versus T at $x = 0.667$. The points marked ES are from the work of Enderby and Simmons (1969) (Donally and Cutler, 1972).

with T. A similar two-band formula applies to the thermopower S. Since S_p and S_n have opposite signs, increasing T and σ_n/σ_p would cause S to change from positive to negative. This has been observed experimentally (Field, 1967).

These results also lead to several conclusions about the possible interpretation of the Hall effect in liquid semiconductors:

(1) The sign discrepancy with S cannot be resolved in a trivial way by a two-band model.

(2) The correlation of μ_H and σ noted by Allgaier is at most a rough statistical one. In Tl_xTe_{1-x}, μ_H decreases as x changes from $\frac{2}{3}$ to 0, whereas σ increases.

(3) As we show later (Section 7.1), the density of electrons in the conduction band can be inferred with confidence from the composition for $x > \frac{2}{3}$. In this case (where S and R_H are both negative), Eq. 2.1 gives a value for n which is at least thirty times too large.

The same sign discrepancy between R_H and S has been observed in several chalcogenide glasses (Pearson, 1964; Male, 1967) and in amorphous germanium (Clark, 1967). The Hall mobility is also independent of T with a magnitude similar to liquids (~ 0.1 cm^2/V sec) even though the resistivity is many orders of magnitude greater and depends very strongly on T. A particularly interesting series of Hall measurements were made by Male (1967) in the glasses As_2Se_2Te, As_2SeTe_2, $As_2Se_3Tl_2Se$, and $As_2Se_3 \cdot Tl_2Te$. As shown in Fig. 2.13, the temperature range encompasses both the glassy state and the liquid state, and the behavior of μ_H is essentially the same over the entire range of T.

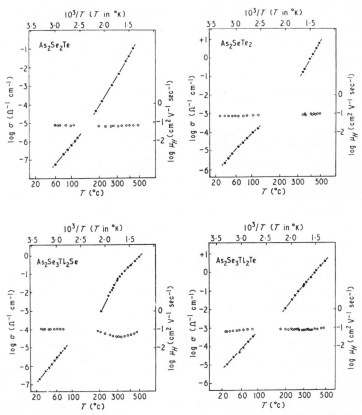

Fig. 2.13. Hall mobility (open circles) and electrical conductivity (solid circles) for several chalcogenide glasses in the vitreous and liquid state (Male, 1967).

2.4 THERMAL CONDUCTIVITY AND THERMOELECTRIC
FIGURE OF MERIT

The thermal conductivity κ has been measured for a number of liquid semiconductors. This work has been reviewed recently (Regel *et al.*, 1971), so our discussion can be abbreviated accordingly.

It should be noted first that accurate measurements of κ are very difficult to make at high temperatures even for solids, and more so for liquids. This is indicated by the frequent appearance of conflicting results. For this reason, it is sometimes risky to make detailed inferences from experimental data. Regel and coworkers (1971) estimate the typical accuracy to be 10–20%.

An exhaustive list of liquids for which κ had been measured, together with references, can be found in the review article (Regel *et al.*, 1971). It includes mostly molten compounds (about 30 of them) as well as a number of alloy systems such as Bi_2Te_3–Bi_2Se_3, Sb_2Se_3 doped with Sb or Te, Tl–Te, Te–Sb, and Te–Se. Most measurements were made over a moderate range of T near the melting temperature.

It is usually convenient to analyze the thermal conductivity in terms of three contributions:

$$\kappa = \kappa_a + \kappa_e + \kappa_r, \qquad (2.3)$$

where κ_a is due to atomic motion, κ_e is due to electronic motion, and κ_r is due to radiation. The quantity κ_a sets a lower limit to κ, and it has a value of 0.5–2.0×10^{-3} cal cm^{-1} deg^{-1} sec^{-1} for most known liquids. This fact can be understood in terms of the expression:

$$\kappa_a = CD, \qquad (2.4)$$

where C is the heat capacity per unit volume and D is the thermal diffusivity. Most of the thermal motion of a liquid near its melting point is vibrational in character, so that the equipartition law gives a value $3k/(4\pi a^3/3)$ for C, where a is the atomic radius and k is the Boltzmann constant. At high temperatures, vibrational energy diffuses with a mean distance $\cong 2a$, so that $D \sim 4a^2v$, where v is the vibration frequency. Putting this together gives $(9/\pi)(kv/a)$ for κ_a, and its small range of values can be ascribed to the fact that a and v for heavier elements do not vary much about the values 2 Å and $\sim 3 \times 10^{12}$ sec^{-1}, respectively.

A number of more refined semiempirical formulas for κ_a exist which are expressed in terms of parameters such as the molecular weight, the melting temperature, and the density. These are based ultimately on the same type of analysis as the one above (Regel et al., 1971). Regel et al. have shown that the other parameters tend also to be correlated with the average atomic weight A, so that very many liquids have values of κ_a which fall on a common curve $\kappa_a (\equiv \kappa_L) = f(A)$. This curve, shown in Fig. 2.14, provides a convenient

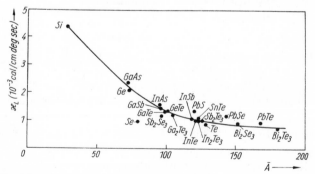

Fig. 2.14.   The calculated atomic thermal conductivity as a function of the average atomic weight at the melting point of various liquids (Regel et al., 1971).

empirical formula since A is generally known. For $A > 100$, which includes most of the liquid semiconductors which have been studied extensively, κ_a is within 10–20% (the usual experimental error) of the value 1×10^{-3} cal/cm deg sec.

For strongly degenerate electronic systems, κ_e is related to the electrical conductivity by the Wiedemann–Franz law:

$$\kappa_e = W_0 \sigma T, \tag{2.5}$$

where the constant $W_0 = (\pi^2/3)(k^2/e^2) = 2.45 \times 10^{-8}$ V²/deg². Measurements of liquid metals show that this relationship is generally obeyed within the typical experimental accuracy, but $\kappa/\sigma T$ is frequently somewhat smaller than the theoretical value. The experimental error can be ascertained to be small enough in several cases to indicate that the discrepancy is probably real (Faber, 1972; Regel *et al.*, 1971). It will be shown (Section 6.3) that the metallic approximation can be expected to be accurate when $\sigma \gtrsim 500$ ohm⁻¹ cm⁻¹, which translates, for typical temperatures ($\sim 1000°$K), to $\kappa_e \gtrsim 3 \times 10^{-3}$ cal deg⁻¹ cm⁻¹ sec⁻¹. Thus, in the range where $\kappa \gg \kappa_a$, the Wiedemann–Franz law describes κ with good accuracy.

Since κ_a can be expected to change relatively slowly with T, a good test of the theory for $\kappa_a + \kappa_e$ is to plot κ versus σT and look for a straight line with slope W. We reproduce such a plot from the work of Perron (1970b) on Te–Se alloys in Fig. 2.15. It shows good agreement with theory, yielding $\kappa_a (\equiv \chi_R) = 3.5 \times 10^{-3}$ W/deg cm and $W (\equiv L) = 2.0 \times 10^{-8}$ V²/deg².

For liquids with smaller electrical conductivities, the theory for κ_e becomes more complicated. For one thing, the value of $W = \kappa_e/\sigma T$ will change when the electron density becomes nondegenerate in single-band transport, approaching in the Maxwell–Boltzmann limit another constant value of similar magnitude. Added transport of thermal energy may also occur when both electrons and holes are present (ambipolar diffusion). The possibility of making inferences about the electronic structure of liquid semiconductors

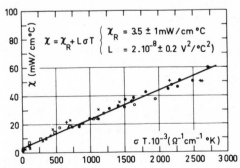

Fig. 2.15. Linear dependence of the thermal conductivity ($\equiv \chi$) on σT in Se_xTe_{1-x} for $x = 0, 0.05, 0.1, 0.2, 0.3,$ and 0.4 (Perron, 1970a).

from ambipolar effects in κ_e has probably motivated some of the interest in measurements of κ. However the theory of this phenomenon is complicated enough so that it is better treated in Section 6.3.

In nonmetallic materials (where κ_e is relatively small) it is possible for radiation to contribute appreciably to heat transport. The contribution of radiation κ_r has been derived theoretically for the case where the optical absorption coefficient α is small compared to the inverse length L^{-1} of the sample:

$$\kappa_r = \frac{16}{3} n^2 \frac{\sigma_b T^3}{\alpha}, \qquad (2.6)$$

where n is the refractive index and σ_b is the Stefan–Boltzmann constant (Genzel, 1953).

This formula assumes that α and N are constant, and it must be modified if they vary appreciably in the range of wavelength where the distribution of thermal radiation peaks. Another complication occurs when $\alpha L \gtrsim 1$, which is apt to be the case if α is small enough to make κ_r large enough to be significant. In this case, the geometry of the sample and its enclosure plays a role since some of the radiation passes through the sample with little or no absorption. This type of situation has been considered by Dixon and Ertl (1971) in a study of liquid Tl_2Te. However, as discussed in Section 7.4, their results seem to be in error because their derived value of α is much too small to be consistent with the electronic transport properties of Tl_2Te.

Part of the interest in measuring the thermal conductivity, as well as the thermoelectric parameters σ and S, has been to determine the potential value of liquid semiconductors in devices for thermoelectric conversion. The figure of merit for thermoelectric conversion is a dimensionless quantity γ (frequently referred to as ZT):

$$\gamma = S^2 \sigma T / \kappa. \qquad (2.7)$$

If $\gamma \sim 1$, an appreciable fraction of the Carnot efficiency can be derived from the thermoelectric conversion process (Ioffe, 1957). For metallic materials, $\kappa \cong \kappa_e$ since κ_a is smaller than κ_e, so that the Wiedemann–Franz law yields $\gamma \cong S^2/W_0$. However for metals, $S \ll k/e$, so that $\gamma \ll 1$ (see Eq. 2.5). The quantity S can be increased by reducing the electron density; doing this also reduces the electrical conductivity, so that κ_e also decreases. But $\sigma T / \kappa$ remains nearly constant as long as κ_e remains large compared to κ_a. In most useful thermoelectric materials, the point of diminishing returns occurs when κ_e becomes comparable to κ_a. The particular interest in liquid semiconductors comes from the fact that κ_a can be expected to be appreciably smaller than in crystalline solids at the same temperature, so that it may be possible to achieve larger values of S before the factor $\sigma T / \kappa$ in γ starts to decrease. Some liquid semiconductors have in fact been found to have values of $\gamma \sim 1$ (see references in Regel *et al.*, 1971); exploitation of these materials has apparently been inhibited so far by practical limitations, such as small ranges of T in which γ is large, and by the engineering and metallurgical problems in using liquids in high-temperature devices.

2.5 MAGNETIC SUSCEPTIBILITY

Most existing information on the magnetic susceptibility χ of liquid semiconductors is derived from melted semiconducting compounds, and many of these results are to be found in the work of Glazov and coworkers (Glazov *et al.*, 1969). Before considering them, it is worthwhile to review a study by Busch and Yuan (1963) of the magnetic susceptibility of molten B elements (post transition elements).

We discuss the theory of magnetic susceptibility in Section 6.4, but for the present purpose we note that for metals, χ can be decomposed into a diamagnetic term χ_{ion} due to the ionic cores and positive χ_{PL} due to the electron gas. Some of the elements discussed by Busch and Yuan (Se, S, I, P) are molecular in both liquid and solid state, and as a consequence they have a diamagnetic susceptibility which does not change appreciably on melting. In the case of simple metals such as Zn or Tl, χ changes very little on melting, and Busch and Yuan have analyzed χ in terms of empirical values of χ_{ion} and the theoretical value of χ_{PL} for a free election gas. The accuracy of this analysis has been questioned (Busch and Güntherodt, 1974), but that need not concern us here, since we are mainly concerned with the implications of the changes in χ on melting various semiconducting solids; this aspect is not sensitive to the questions which have been raised.

Crystalline semiconductors such as Ge and Si have negative values of χ which are in a range characteristic of insulators, but on melting the change $\Delta\chi$ is large and positive, and the final values are in approximate agreement with $\chi_{ion} + \chi_{PL}$ expected for metals. The behavior of χ on melting Ge is shown in Fig. 2.16a. The same type of behavior occurs when semimetals such as Sb and Bi are melted. In the case of InSb or Te, Busch and Yuan found that although $\Delta\chi$ is large and positive, the final value falls short of the value for $\chi_{PL} + \chi_{ion}$ expected for a metal. The behavior of Te, shown in Fig. 2.16b, is particularly interesting because χ increases further as the temperature of the liquid is raised above the melting point, suggesting that the liquid is becoming more metallic. The change is qualitatively similar to the change in the electrical conductivity (see Fig. 2.5), and both $\sigma(T)$ and $\chi(T)$ tend to level off at high T.

We turn now to the results of Glazov *et al.* (1969), who studied systematically the behavior of $\Delta\chi$ on melting a number of crystalline semiconductors.

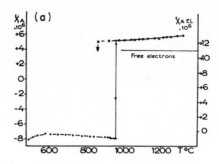

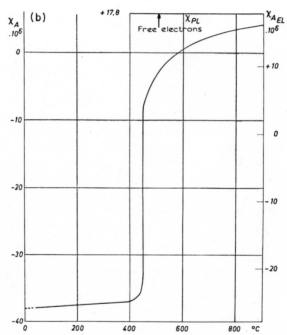

Fig. 2.16.   Behavior of the magnetic susceptibility on melting (a) Ge (b) Te (Urbain and Übelacker, 1967).

For Ge, Si, and the III–V compounds AlSb, InSb, GaAs, and InAs, they observed that $\Delta\chi$ is positive and $d\chi/dT$ is small for the liquid. This indicates a transition from covalent to metal bonding, and the electronic structure in the liquid state changes little with T. This is in accord with the behavior of σ (see Table 2.1), which is in the metallic range ($>10^4$ ohm^{-1} cm^{-1}) and has a small temperature coefficient in the liquid state (Glazov, 1969).

A different behavior is observed on melting ZnTe, CdTe, In_2Te_3, and Ga_2Te_3. In addition to a positive $\Delta\chi$ on melting, there is a further rise with T in the liquid state such as occurs in molten Te in Fig. 2.16b. In every case, there is a parallel increase in $\sigma(T)$ (Glazov, 1969), and as one can see in Table 2.1, $\sigma_{liq} < 100$ ohm^{-1} cm^{-1} at the melting point. (It is considerably larger for tellurium.) The inference is that considerable covalent bonding persists in the molten compound, and a further change toward metallic behavior results from increasing T. The compounds PbTe, GeTe, and SnTe are similar to the preceding compounds but do not show a clearcut behavior of $\chi(T)$ above the melting point. Their electrical conductivities fall in a range (1000–3000 ohm^{-1} cm^{-1}) which is not fully metallic (Table 2.1).

Relatively few binary alloys have been studied which form semiconducting liquids. These include Tl–Te (Brown et al., 1971; Gardner and Cutler, 1976) and Cs–Au (Steinleitner and Freyland, 1975), and their implications will be discussed in Chapters 7 and 8.

CHAPTER 3

PHYSICOCHEMICAL AND
METALLURGICAL PROPERTIES

In common with all condensed matter, the electronic behavior of liquid semiconductors is a reflection of the physical structure on an atomic scale. An important part of our problem is to understand as much as possible about this structure, and physicochemical data are an important source of information for this purpose. Chemical factors also determine the experimental range of composition and temperature in which one can make measurements of liquid semiconductor behavior, and this is a good place to review these factors.

3.1 THE CHEMICAL RANGE OF LIQUID
SEMICONDUCTOR SYSTEMS

Aside from the elements Se and Te, liquid semiconductors are alloys containing two or more elements. Tertiary or more complicated alloy systems can be understood by natural extensions of considerations concerning binary alloy systems, so the discussion will be confined to binary systems.

It is a well-known metallurgical fact that crystalline intermetallic compounds which have covalent or ionic bonds tend to form when at least one of the elements is a member of the IVB or later group of the periodic table (Hume-Rothery and Raynor, 1962, Chapter 4). This follows from the chemical principle that compound formation is favored when one of the elements has a nearly filled valence electron shell. It is evident that the chemical compositions of liquid semiconductors are in accord with this principle. The theory of Mooser and Pearson (Mooser and Pearson, 1960) on the chemistry of bonding of semiconductor compounds is a more detailed exposition of this principle. The main difference for liquid semiconductors is that the liquid phase can accommodate a larger variety of molecular structures than crystals, so that the stoichiometric limitations according to the rules of Mooser and Pearson cannot be applied strictly.

Having noted this, we consider why semiconductor behavior seems to be confined largely to alloys containing a chalcogenide (VIB) element, and the role of the electronegative element is rarely taken by IVB, VB, or VIIB elements. Part of the answer lies in the fact that IVB and VB elements, particularly the heavier ones, have a strong tendency to assume an alternate valence which is smaller by two. This corresponds to not using the lower $(ns)^2$ valence electrons in bonding. Thus elements such as tin or antimony have an alternate chemical behavior similar to cadmium or indium, respectively, and tend to form typical metallic alloys when combined with other metals. As for the group VIIB elements, their large electronegativity causes them to form ionic compounds. However, a number of alloy systems M–X, where X is a halogen, exist over appreciable ranges of stoichiometry. These systems have many of the characteristics of liquid semiconductors as noted in Chapter 1.

A second limitation to the range of liquid semiconductors is the formation of a solid or vapor phase. It should be appreciated that the occurrence of a particular phase reflects only a greater stability relative to alternative phases. The possibility of forming a particularly stable solid compound limits the temperature range of the liquid from below. Elements which form small stable molecules in the vapor phase have an especially large vapor pressure, and this can limit the practical temperature range of the liquid from above.

An extensive discussion of stability of solid compounds is beyond the scope of our work, but, as elucidated in a very informative treatise by Wells (1962, 1958), it is governed by a combination of factors relating to the bond type, strength of the bonds, valence combinations, atomic size, and crystal packing, which determine the relative stability of various possible crystal structures. As a result of these factors, for instance, binary systems which satisfy chemical valence in bonding to form a compound AB, such as CdTe or SnSe will have a relatively stable solid compound at this composition which melts at a high temperature, and which remains in equilibrium with a liquid phase at other compositions at relatively high temperatures. Thus the liquid phase occurs only at temperatures above 900°C near these compositions. Binary compounds of the valence type III–VI do not seem to readily form stable crystal lattices, and their melting points tend to be relatively low.

With regard to the high T limitations, the elements O, S, Se, and Te form relatively stable gaseous molecules. Thus liquid chalcogenide alloy systems are limited in their practical temperature range by high vapor pressures, particularly for the first three elements, at compositions rich in the chalcogen.

The compilation of phase diagrams and other information on phase equilibrium for binary systems (Hansen and Anderkov, 1958) is a very useful reference on the range of composition and temperature for liquid semiconductors. The considerations in the preceding paragraphs serve well to

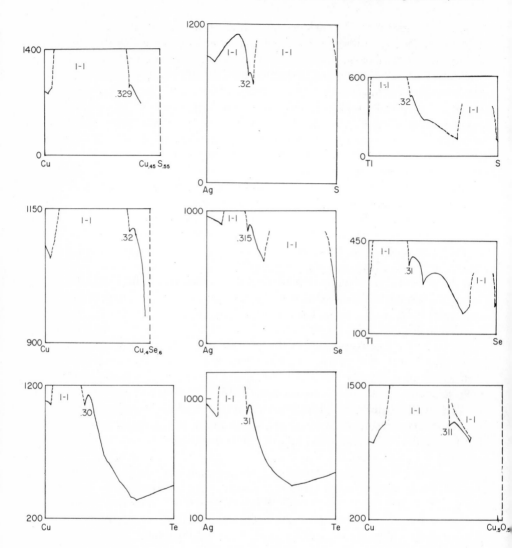

explain, if not predict approximately, the extent of this range for various systems. We show in Fig. 3.1 the liquidus curves for a number of binary systems which form liquid semiconductors.

Composition ranges in which two immiscible liquid phases separate occur frequently in phase diagrams of liquid semiconductor systems. It is particularly interesting in view of the chemical principle that dissimilar liquids do not mix. The separation, for example, of Tl–Te (Fig. 2.4) into a phase with composition of approximately 100% Tl and another with approximately

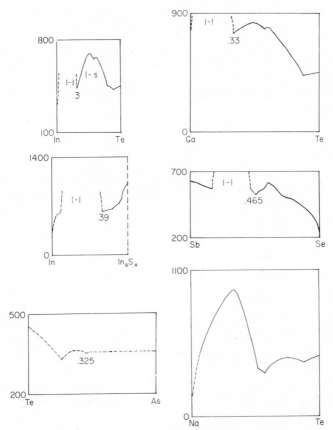

Fig. 3.1. Simplified phase diagrams for a number of binary alloys of interest as liquid semiconductors. Liquid–liquid phase separation occurs below dashed lines, and liquid–solid phase separation occurs below solid lines. Temperatures are in °C.

70% Tl suggests that the latter is a semiconductor which prefers not to "mix" with the metallic Tl. The composition of the "pure" liquids are only approximately determined by this consideration, since immiscible liquids always have some solubility in each other.

The apparent separation between a "metallic" and a "semiconductor" phase of composition M_2X occurs in a number of binary systems, as seen in Fig. 3.1: Tl–Te, Tl–Se, In–Te, Ag–S, Ag–Se, Ag–Te, Cu–O, and Tl–S. Another frequently occurring type of miscibility gap is found in alloys rich in Se or S. Here a "semiconductor" phase separates from a liquid S or Se, the latter apparently being a molecular liquid with van der Waal binding between the molecules. (Se is commonly regarded as a semiconducting liquid, but its electronic behavior is near the insulator extreme, like sulfur.) Of course, immiscibility does not necessarily mean that one of the phases is a semiconductor. Many liquids separate in which both constituents are clearly metallic.

3.2 DENSITY

The connection has been noted earlier between the atomic volume and the nature of binding which suggests that many liquid semiconductors have a covalent structure. This is based on the fact that covalent bonds generally require a relatively small coordination number z (number of nearest neighbors) which is typically 4 or 6. In metallic binding, it is approximately 12 in solids and 9 in liquids, reflecting the geometry of closely packed spheres in crystals and randomly packed spheres in disordered materials.

The volume per atom is determined by both the coordination number and the distance between atoms. The latter is generally determined from diffraction studies, as discussed in Section 3.5, but it is generally true that the interatomic spacing increases with temperature and on melting. Consequently, when the density increases on melting or with increasing temperature, one can safely infer from the density behavior alone that z is increasing. On the other hand, an increase in z with temperature or fusion may be obscured by a larger opposing effect due to an increasing interatomic distance. The most famous example of an increasing z occurs in water, where the density increases both on melting and with increasing T up to $4°C$.

Ioffe and Regel (1960) have examined cases where the density of a semiconductor increases on melting, and have correlated the inferred increase in z with a transition to metallic behavior. This was observed in Ge, Si, and some III–V semiconducting compounds (see Table 2.1). In the case of Te and dilute alloys of Se in Te, the volume increases on melting, but there is a range of increasing density above the melting point as shown in Fig. 3.2. It should be noted that the occurrence of density maxima in these alloys has been disputed (Urbain and Übelacker, 1969; Lucas and Urbain, 1962), but it has been confirmed in more recent work (Ruska, 1973). The electrical conductivity increases both on melting and further heating, reflecting a tendency toward metallic behavior. The inference is that z increases in both processes, but on melting, the increasing interatomic distance (mainly between chains of the stacked chain structure of the solid) causes a net increase in volume. More will be said about the structure of liquid Te in the succeeding sections.

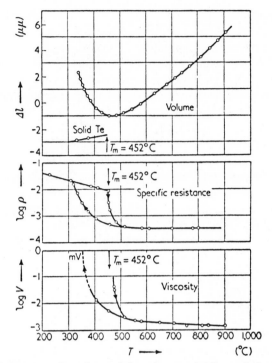

Fig. 3.2. Temperature dependence of the volume, resistivity, and viscosity of tellurium (Ioffe and Regel, 1960).

Much more extensive data have been obtained by Glazov and coworkers (Glazov *et al.*, 1969). They found an increase in density on fusion in III–V semiconductor compounds such as InSb, GaSb, GaAs, and compounds of the type Mg_2Sn (see Table 2.1). In all of these, the melts are metallic. For other substances, such as Ga_2Te_3, In_2Te_3, Se, GeTe, Bi_2Se_3, where the melts are semiconductors or nearly so, the density decreases on melting. For the first two materials just mentioned, there is a range of T above the melting point in which the density increases as in Te. This is consistent with the view that although z increases in all cases on melting and with increasing T, the increase is less complete for materials which are semiconductors in the liquid state than for materials which become metallic.

Density measurements of liquid semiconductors can be of value in two other ways. One is to convert information about composition from units of mass into units of volume. This is needed for theoretical interpretation of experimental information. However, since the volume changes on melting and thermal expansion are relatively small, it is frequently sufficient to use the density of the corresponding solid or a weighted average of the atomic volumes of the pure liquid constituents of the alloy. Tables of atomic volumes of pure liquid metals have been published by Wilson (1965).

One can also find evidence of compound formation from a plot of the atomic volume versus composition for a binary alloy. We show such a plot for Tl–Te in Fig. 3.3. In this case, the excess volume (measured from a straight line connecting the values for the pure elements) has a broad maximum in the vicinity of the composition Tl_2Te (Dahl, 1969; Nakamura and Shimoji, 1973). Similar broad peaks occur in In_xTe_{1-x} and Ga_xTe_{1-x} near $x = 0.4$ (Lee, 1971). Usually the excess volume is less than $\sim 10\%$, which indicates the accuracy of the estimate suggested in the preceding paragraph.

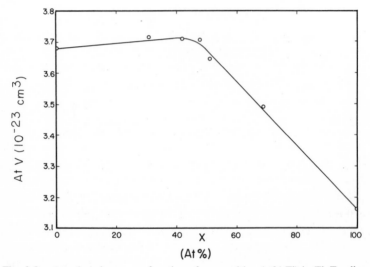

Fig. 3.3.   Atomic volume as a function of composition (at% Tl) in Tl–Te alloys.

3.3   VISCOSITY

The viscosity of a liquid is expected to be very sensitive to the microscopic structure. Although the theory of this relationship is very difficult, so that the exact interpretation of experimental results is nebulous, significant conclusions can be obtained about some liquid semiconductors on the basis of broad considerations.

An extensive discussion of the viscosity of metallic and molecular liquids has been presented by Egelstaff (1967). The viscosities of liquid metals generally lie in the range $0.5–5 \times 10^{-2}$ poise, and Egelstaff notes that molecular liquids generally fall in a similar range. Glazov and collaborators have made viscosity measurements of many liquid semiconductors, and these also tend to fall in this range (Glazov et al., 1969). Liquids such as long chain organic polymers have viscosities which are larger by orders of magnitude.

The behavior of the viscosity η of Te (Ioffe and Regel, 1960) provides an interesting example of the possible relationship of η to formation of chain molecules. Figure 3.2 shows the concurrent behavior of η, d, and σ of molten Te. As noted earlier, the behavior of d suggests that the stacked chain structure of crystalline Te only partially disintegrates on melting, yielding a mixture of chain molecules whose average length decreases further with temperature. The behavior of η is completely in accord with this model. The quantity σ increases as η decreases, which suggests that metallic behavior is associated with an increased density of broken bonds. A different interpretation of the structure of liquid Te has been derived from neutron diffraction studies; this interpretation is discussed in Section 3.5. Crystalline Se also has a stacked chain structure, and the behavior of η on melting is qualitatively similar to that of Te. The viscosity is much larger, and η decreases with increasing concentration of Te in Te–Se alloys. In the case of Se, various kinds of evidence discussed in Section 8.6.1 confirm the chain structure of the liquid phase.

There are a number of alloy systems which readily form glasses on cooling. In these alloys, the viscosity is also very large, and the enormous increase in η on cooling provides the mechanism which inhibits the nucleation and growth of the more stable crystalline phases. These systems are typically tertiary alloys which are chemically similar to As_2Se_3. Since they contain elements such as As or Sb which form three or more covalent bonds, the molecular linkages create networks rather than chains. As indicated by the low values of σ, the molecular structure persists over a large range of temperature above the melting temperature (Male, 1967), in contrast to the behavior of Te. This conclusion is also supported by studies of the optical properties (Taylor et al., 1971).

Very extensive studies of the viscosities of liquid semiconductors have been carried out by Glazov and coworkers (1969). These include investigations of both melts of solid compounds, such as SnTe, ZnTe, and In_2Te_3, and binary alloys in which the composition is a variable. Generally, η is in the range typical of molecular and metallic liquids, in contrast to the examples discussed so far.

The $\eta(T)$ curves for molten compounds were analyzed by Glazov and coworkers in terms of the behavior of the activation energy E_a as the temperature is increased above the melting point. In many cases they found changes in E_a between low temperature and high temperature ranges, which they have interpreted as evidence of the development of molecular structure at the lower temperatures. Generally, η decreases monotonically with increasing T, but maxima were observed in Ga_2Te_3, In_2Te_3, and in binary alloys at neighboring compositions.

Conclusions about structure which are derived from changes in the activation energy of $\eta(T)$ may be open to some question, since they are based on a particular theory for viscous flow due to Eyring (Glasstone *et al.*, 1941). This is one of a number of theories which has been proposed for the viscosity. The proper model for interpreting η in any detail is still an open question. However the existence of molecular structures seems to be indicated more directly by the character of the viscosity isotherms which were determined in a number of binary alloy systems.

Viscosity isotherms for Pb–Te are shown in Fig. 3.4 together with the corresponding electrical conductivity isotherms. It is seen that the appearance

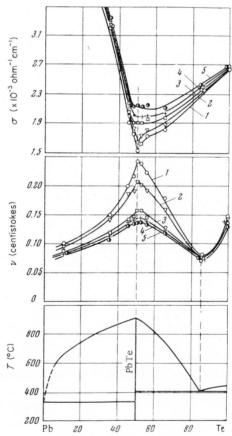

Fig. 3.4. Viscosity and electrical conductivity isotherms and equilibrium phase diagram of Pb–Te. Temperature (°C): (1) 950, (2) 1000, (3) 1100, (4) 1150, (5) 1200 (Glazov *et al.*, 1969).

of a singular composition, at which η has a peak and where η is most sensitive to temperature, mirrors a similar behavior in the resistivity. This suggests that chemical bonding is responsible for both. There is a similar correspondence in a large number of other binary systems studied by Glazov *et al.*, and these results are summarized in Table 3.1.[†]

In addition to the binary systems studied by Glazov *et al.*, viscosity measurements have been reported for Tl_xTe_{1-x} (Nakamura and Shimoji, 1973) and Tl_xSe_{1-x} (Kazandzhan, *et al.*, 1974). In each case, both η and $-d\eta/dT$ peak at the composition $x = \frac{2}{3}$ where singularities occur in the thermoelectric properties.

TABLE 3.1

Viscosity Behavior of Binary Alloy Systems[a]

M_xA_{1-x}	Composition of viscosity maximum	M_xA_{1-x}	Composition of viscosity maximum	M_xA_{1-x}	Composition of viscosity maximum
Al–Sb	0.5	Cd–Te	0.5	Bi–Se	0.4
Ga–Sb	0.5	Ga–Te	0.5, 0.4	Bi–Te	0.5, 0.4
In–Sb	0.5	In–Te	0.5, 0.4	Sb–Te	0.4
Mg–Pb	0.67	Ge–Te	0.5	Tl–Te[b]	0.67
Mg–Sn	0.67	Sn–Te	0.5	Tl–Se[c]	0.67
Zn–Te	0.5	Pb–Te	0.5		

[a] Unless otherwise noted, data are from Glazov *et al.* (1969).
[b] Nakamura and Shimoji (1973).
[c] Kazandzhan *et al.* (1974).

3.4 THERMOCHEMICAL INFORMATION

We have noted a number of indications that molecular structure plays a key role in liquid semiconductors. Thermochemical data provide a powerful means for inferring the existence of compounds in binary alloys, and therefore are a potential source of information about the structure. Before discussing this in detail, it is worth noting that there is a large body of semi-empirical information concerning bonding between elements (Pauling, 1960)

[†] The determinations of η by Glazov and co-workers were made simultaneously with σ by the measurement of the amplitude and attenuation rate of a torsional oscillator immersed in the liquid (see Section 4.2). The numerical values of σ and η are extracted from the observed parameters by means of theoretical equations for the method. In such a situation, there is a danger that experimental errors would cause singularities in the behavior of one of the parameters to appear spuriously in the other. We suspect that this factor accounts for the discrepancies between their results for σ and those of other investigators such as occur in In–Te (Blakeway, 1969; Zhuze and Shelykh, 1965; Ninomiya *et al.*, 1973).

which is derived from thermochemical as well as other sources. This can often be used to make estimates of bond energies and molecular geometry in the absence of more direct experimental information.

The experimental and the theoretical methods which lead to information about the enthalpy H, entropy S, and the free energy (Gibbs potential) G are described in standard treatises (Lewis and Randall, 1961). Treatises on the theory of solutions include those of Guggenheim (1952) and of Hildebrand and Scott (1950). The thermochemical study of metallic alloys is particularly relevant to liquid semiconductors, and the reader is referred to a review by Kubaschewski and Evans (1958).

Experimental data for binary alloys $A_x B_{1-x}$ are frequently reported in terms of the enthalpy of mixing ΔH_m, the entropy of mixing ΔS_m, and the free energy of mixing ΔG_m, and this information is often in the form of an isotherm as function of the composition parameter x. (In this section, we shall always refer to molar quantities.) These parameters are simply the excess over what would be expected if the constituents did not interact, and they are expressed graphically by the deviation of the isotherm from a straight line connecting the values for the pure constituents. For instance:

$$\Delta G_m(x) = G(x) - (1 - x)G(0) - xG(1). \tag{3.1}$$

There is a useful relationship between G and the chemical potentials μ_A and μ_B of the constituents:

$$G(x) = x\mu_A + (1 - x)\mu_B, \tag{3.2}$$

where:

$$\mu_A = -\mu_B = (\partial G/\partial x)_{T,P}.$$

This leads to a relationship for ΔG_m which will be used later:

$$\Delta G_m(x) = x[\mu_A(x) - \mu_A(1)] + (1 - x)[\mu_B(x) - \mu_B(0)]. \tag{3.3}$$

For an ideal solution, ΔS_m has the form:

$$\Delta S_I = -R[x \ln x + (1 - x) \ln(1 - x)]. \tag{3.4}$$

Although many alloys deviate appreciably from ideal mixtures, they often are qualitatively in agreement with Eq. 3.4 in that ΔS_m is positive with a maximum value $\sim R$ at $x \sim 0.5$. The quantity ΔH_m is effectively the energy of interaction between A–B neighboring atoms, as compared to A–A neighbors and B–B neighbors. For weak interactions between A and B, ΔH_m is comparable to the thermal energy RT. In this case one can derive a theoretical form (Guggenheim, 1952):

$$\Delta H_m = Wx(1 - x), \tag{3.5}$$

where the excess interaction energy W is positive or negative, but is generally less than RT in magnitude. Again, real alloys usually differ from this in detail, but agree qualitatively. We show in Fig. 3.5 curves for ΔH_m and ΔS_m for a number of binary alloys from a paper by Terpilowski and Zaleska (1963) which illustrate this. The quantity ΔS^{ex} which is plotted here is $\Delta S_m - \Delta S_I$.

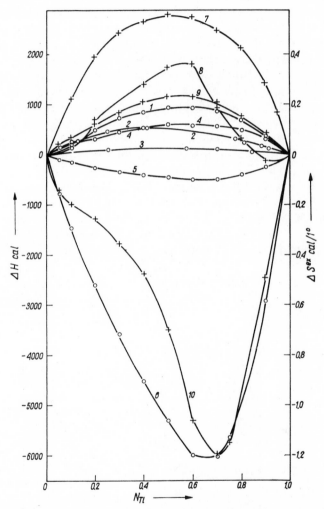

Fig. 3.5.   Isotherms for the enthalpy of mixing (ΔH) ($\bigcirc$) and the entropy of mixing in excess of the value for ideal mixtures (ΔS^{ex}) ($+$) for a number of liquid binary alloys as a function of the atomic fraction of Tl, N_{Tl}. ΔH: (1) Ag–Tl, 777°C; (2) Cd–Tl, 450°C; (3) In–Tl, 350°C; (4) Sn–Tl, 478°C; (5) Sb–Tl, 650°C; (6) Te–Tl, 600°C. ΔS^{ex}: (7) Ag–Tl, 777°C; (8) Sn–Tl, 478°C; (9) Sb–Tl, 650°C; (10) Te–Tl, 600°C. (Terpilowski and Zaleska, 1963.)

Figure 3.5 also illustrates the distinctly different behavior of Tl–Te which strongly implies that a compound forms. The quantity ΔS^{ex} is strongly negative (in units of R). This implies an ordering process which may reflect the formation of molecules. The quantity ΔH_m also has a negative peak whose magnitude is large compared to RT. The magnitude 6.0 kcal/g atom implies an enthalpy of formation of ~ 18 kcal/mole for Tl_2Te.

In addition to the measurements by Terpilowski and Zaleska, thermochemical studies of Tl_xTe_{1-x} have been made by Nakamura and Shimoji (1971) (ΔG_m, ΔH_m, ΔS_m), by Castanet et al. (1968) (ΔH_m), and by Maekawa et al. (1971a) (ΔH_m). The results of Nakamura and Shimoji seem to be the most accurate, and we show their result for ΔG_m in Fig. 3.6. The fact that the minimum near $x = \frac{2}{3}$ is several times RT in magnitude implies that the dissociation constant for the reaction $Tl_2Te = 2Tl + Te$ is relatively small. There is also thermochemical information on ΔG_m, ΔH_m and ΔS_m for the binary alloys Tl–Se (Terpilowski and Zaleska, 1965), Tl–S (Fukuda et al., 1972) and Bi–Te (Liu and Angus, 1969). In addition there have been measurements of ΔH_m in Bi–Te (Egan, 1959; Maekawa et al., 1971a), Tl–Se and Tl–S (Maekawa et al., 1971b), In–Te and Sb–Te (Maekawa et al., 1972a), and Bi–Se and Sb–Se (Maekawa et al., 1972b). Although the composition range is incomplete in some cases (Tl–Se, Tl–S), the isothermal curves are always consistent with compound formation at the composition where the resistivity is a maximum. It should be noted that many binary liquid alloys exist for which there is similar thermochemical evidence for compound formation, but whose electrical conductivities remain in the range ($\gtrsim 10^4$ ohm^{-1} cm^{-1}) which we classify as liquid metals rather than liquid semiconductors. Examples of these are KTl, Mg_2Pb, and KHg_2.

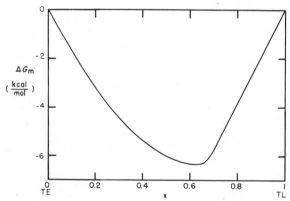

Fig. 3.6.   The Gibbs free energy of mixing for Tl_xTe_{1-x} at 600°C (from the data of Nakamura and Shimoji, 1971).

Is it possible to derive more quantitative information about the structure of the liquid? This may be done in two ways. One is to set up a thermodynamic model based on the theory of solutions and investigate whether it agrees with experiment for reasonable values of the model parameters. A second approach is to analyse the ΔG_m curves in terms of statistical mechanical theory for fluctuations.

Interesting results have been obtained from solution models by two different approaches. Nakamura and Shimoji (1971) treat $Tl_x Te_{1-x}$ as a pseudobinary solution of $Te + Tl_2 Te$ for $x < \frac{2}{3}$ and $Tl_2 Te + Tl$ for $x > \frac{2}{3}$, and compare the entropy of mixing with the theoretical formula. In view of the large difference in volume between the two constitutents, they use a formula due to Flory (1942) which replaces the mole fraction by the volume fraction ϕ in the arguments of the logarithmic terms in Eq. 3.4. The pseudobinary entropy of mixing $\Delta S_m{}^*$ is:

$$\Delta S_m{}^* = -(1 + 2x_2)^{-1} R[x_2 \ln \phi + (1 - x_2) \ln (1 - \phi)]. \qquad (3.6)$$

The quantity x_2 is the mole fraction of $Tl_2 Te$, and ϕ is calculated on the assumption that $Tl_2 Te$ has three times the volume of Tl or Te. The factor $(1 + 2x_2)^{-1}$ coverts the formula to a per-atom basis, for comparison with the experimental data. We show these results in Fig. 3.7. The quantity $\Delta S_m{}^*$ is defined as the height of ΔS_m above secants drawn between compositions tellurium and $Tl_2 Te$ and between $Tl_2 Te$ and Tl. It is seen that there is good agreement with the theoretical dashed curve for $Tl_2 Te + Tl$, and poorer agreement for $Tl_2 Te + Te$. This supports the view that the Tl-rich alloy is a mixture of Tl and $Tl_2 Te$. It also suggests that the Te-rich alloy may not be a simple mixture of $Tl_2 Te$ and atomic Te.

Bhatia and Hargrove (1973) use a more elaborate model in which three species occur in the liquid at all compositions, $Tl_2 Te$, Tl, and Te, and each is described by a chemical potential μ_i derived from the theory of solutions. Since there is chemical equilibrium between the three species, the three chemical potentials are related by the law of mass action. The reader is referred to the original paper for details, and we note only that the model effectively uses Eq. 3.4 rather than Eq. 3.6 for the entropy, and the energy of interaction is a more elaborate version of Eq. 3.5, with three parameters W_{ij} which refer to the three types of pairs of species. Consequently, the model has a large number of parameters: W_{ij}, dW_{ij}/dT, g, and dg/dT, where g is the standard free energy change for the reaction $Tl_2 Te = 2Tl + Te$. A set of parameters was found which reproduces quite well the experimental curves for ΔG_m, ΔS_m, and ΔH_m. It is interesting to note that the value of g determined for $Tl_2 Te$ was $10.8 RT$, which implies that there is very little dissociation of the molecules. Bhatia and Hargrove also applied their theory to

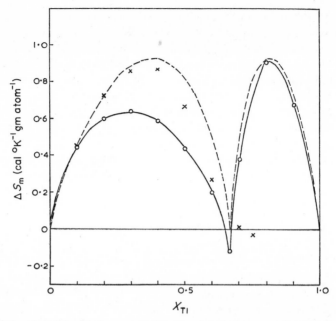

Fig. 3.7. $\Delta S_m^*(x)$ for $Tl_x Te_{1-x}$ at 600°C. The dashed curves are theoretical curves for ideal mixing according to Eq. 3.6. The crosses are experimental points of Terpilowski and Zaleska (1963), and the circles are experimental points of Nakamura and Shimoji (Nakamura and Shimoji, 1971).

Mg–Bi (assuming a third species Mg_3Bi_2), again with good agreement with experiment.

A second way to make quantitative inferences is to use the statistical mechanical theory of fluctuations (Tolman, 1938; Landau and Lifshitz, 1958). A well-known result of this theory is that the mean square of the concentration fluctuation in a binary system $A_x B_{1-x}$ is given by:

$$\langle \Delta x^2 \rangle = RT/N(\partial^2 G/\partial x^2)_T, \tag{3.7}$$

where N is the number of atoms in the region being considered. Thus $(\partial^2 G/\partial x^2)_T$, which is the curvature of $\Delta G_m(x)$, is an inverse measure of the tendency for large fluctuations to occur. Fluctuations have been discussed by various authors in terms of different quantities derived from the above relation. Darken (1967) calls $(\partial^2 G/\partial x^2)_T$ the excess stability and points out that it has sharp peaks at compositions corresponding to alloys which form compounds. In their study of Tl–Te, Nakamura and Shimoji have noted the large peaks in the excess stability at the composition Tl_2Te. Bhatia and Thornton (1970) have defined a number of correlation coefficients in wave-vector space which we discuss in Section 3.6. One of these, the concentration correlation function, $S_{CC}(Q)$ is equal to $N\langle \Delta x^2 \rangle$ in the limit that the wave vector $Q \to 0$. Ichikawa and Thompson (1973) and Bhatia and Hargrove (1974) have discussed fluctuations of a number of binary systems, including Mg–Bi and Tl–Te, in terms of $S_{CC}(0)$.

Aside from the fact that fluctuations are particularly small at the compound compositions Mg_3Bi_2 and Tl_2Te, as indicated by the high curvature of $G_m(x)$, there has been a special interest in the magnitude of fluctuations at other compositions. This interest has been stimulated by the proposal by several authors of theories for the electrical properties of liquid semiconductors based on hetergeneous transport (Hodgkinson, 1971, 1973; Cohen and Sak, 1972; Cohen and Jortner, 1973). They have suggested that relatively large clusters of the compound (say Tl_2Te) or metal (say Te or Tl, depending on which is in excess) are formed at all compositions. These clusters are assumed to be large enough (about 30 atoms are required) so that the overall electrical properties are those of a heterogeneous mixture. With $N = 30$ for the alloys mentioned, $\langle \Delta x^2 \rangle^{1/2}$ is less than 0.1 for all compositions except for Tl_xTe_{1-x} in the region $x > \frac{2}{3}$. Since there is a liquid-liquid phase separation with a critical point at $x \cong 0.8$ and $T = 580°C$, large fluctuations are not unexpected at these compositions. This incipient phase separation causes the very flat shape of ΔG_m in the curve in Fig. 3.6. There are several controversial elements in this theory. As far as the thermodynamic evidence is concerned, a difficulty arises in that the theory for Eq. 3.7 describes a limited aspect of the fluctuations, whereas the heterogeneous transport hypothesis involves large values of both N and Δx. Turner (1973) has approached the problem by calculating the value of N in clusters in which $\langle \Delta x^2 \rangle^{1/2}$ has the required value, and found that it is much too small (much less than 10) for all compositions with the exception noted for Tl_xTe_{1-x} at $x > 0.7$.

A more direct approach to the problem can be made using an expression derived from a grand canonical ensemble for binary alloys (Cutler, 1976c). It is found that the relative probability f of a cluster of N atoms of composition x in a binary system A_xB_{1-x} whose components have chemical potentials μ_A and μ_B is:

$$f = \exp[-(N/RT)\{G(x) - x\mu_A - (1 - x)\mu_B\}]. \qquad (3.8)$$

This expression does not depend on any assumption that x is near the values at which μ_A and μ_B are determined, i.e., the average composition x_0. There are very interesting graphical constructions for the relative probability in terms of:

$$f = \exp[-N \, \Delta G/RT], \qquad (3.9)$$

in which ΔG is obtained by a graphical construction such as is shown in Fig. 3.8. For a fluctuation of N atoms with composition x out of a large volume of average composition x_0, ΔG is the distance between $\Delta G_m(x)$ and a point at x on a tangent to the curve constructed at x_0, as shown in·Fig. 3.8a. For a separation of N atoms at average composition x_0 into two clusters with compositions x_1 and x_2, the geometric construction is Maxwell's construction, so that ΔG is the distance of a point at x_0 on a secant above $\Delta G_m(x_0)$, where the secant is drawn between $\Delta G_m(x_1)$ and $\Delta G_m(x_2)$ as shown in Fig. 8b. Figure 3.9 shows the relative probabilities f_{20} of the separation of a cluster at a single composition x for various values of x_0 in Tl–Te for $N = 20$. Not surprisingly, f becomes extremely small for relatively small values of $(x - x_0)$, except for the near critical compositions where $x_0 > 0.7$. But even there, f is much the same for various values of x. Therefore the fraction of the volume in which the composition is Tl_2Te or Tl is still much less than 1, since intermediate compositions are about equally probable. This point has been overlooked by several authors who indicate that heterogeneous transport may occur in Tl–rich Tl–Te.

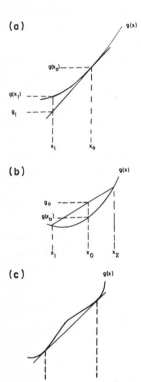

Fig. 3.8.   Graphical constructions for ΔG from the isothermal dependence of G or ΔG_m on x. (a) Fluctuation in which N atoms of composition x_1 separate from N_0 ($\gg N$) atoms of composition x_0. $\Delta G = N[g(x_1) - g_1]$. (b) Maxwell construction for N atoms of composition x_0 separating into N_1 atoms at x_1 and N_2 atoms at x_2. $\Delta G = N[g_0 - g(x_0)]$. (c) Situation where Maxwell's construction yields $\Delta G < 0$. At equilibrium, there is a phase separation between compositions x_a and x_b.

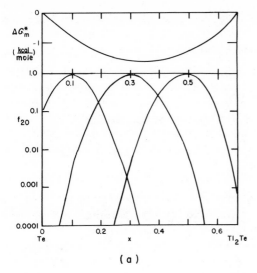

(a)

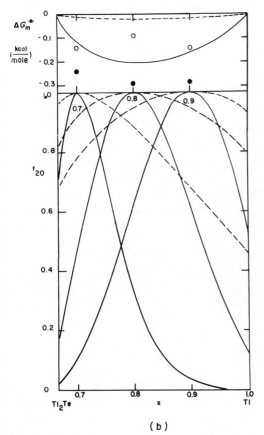

(b)

Fig. 3.9. The relative probability f_{20} for separation of 20 atoms of composition x from an alloy of composition x_0. The value of x_0 is indicated at the peak of each curve. The top of each figure shows ΔG_m^* for the pseudobinary mixture. (a) Pseudobinary mixture Te + Tl$_2$Te at 600°C. (b) Pseudobinary Tl$_2$Te + Tl. The dashed curves and open circles are for $T = 600°$C, and the solid curves and closed circles are for $T = 800°$C.

49

3.5 X-RAY AND NEUTRON DIFFRACTION

Diffraction studies provide, in principle, the most direct way to obtain structural information about liquids. Relatively little has been done with liquid semiconductors, but a discussion of the possibilities of this method of investigation is worthwhile. More complete treatments can be found in work by Egelstaff (1967), Wagner (Beer, 1972), and by Faber (1972), and Enderby (Tauc 1974) has presented a discussion which is directed particularly towards liquid semiconductors.

The basic quantity which is determined from a diffraction measurement is the structure factor $S(Q)$. For a liquid with one type of atom, the scattering intensity for a change in the wave vector Q is proportional to a quantity which is closely related to $S(Q)$. One can derive from $S(Q)$ the pair distribution function $g(r)$. The probability of finding a second particle in a volume element $d\Omega$ at a distance r from the first particle is $(N/\Omega)g(r)\,d\Omega$, where N/Ω is the average density of particles. $S(Q) - 1$ and $g(r) - 1$ are related by three-dimensional Fourier transformations:

$$S(Q) - 1 = \frac{4\pi\Omega}{N} \int_0^\infty [g(r) - 1] \frac{\sin Qr}{Qr} r^2 \, dr, \tag{3.10}$$

$$g(r) - 1 = \frac{\Omega}{2\pi^2 N} \int_0^\infty [S(Q) - 1] \frac{\sin Qr}{Qr} Q^2 \, dQ. \tag{3.11}$$

In determining $S(Q)$, a number of corrections need to be made which are discussed in detail in Enderbys review (Tauc, 1974). Also, experimental limitations lead to a maximum in the available range of Q, so that the calculation of $g(r)$ requires a "termination correction" in the integral. The resulting curves are quite sensitive to these corrections. Consequently, it is not surprising that conflicting results have come from different laboratories, and disagreements about the structure of particular liquids are common in this field.

Bearing these difficulties in mind, let us consider the experimental results that Tourand and Breuil (1970) obtained for molten tellurium by neutron diffraction. Their curves for $g(r)$ at a number of temperatures are shown in Fig. 3.10. The positions of peaks which occur at smaller r indicate the distances of the nearer neighbors of an atom, and their areas indicate the average number of nearest neighbors in each shell. The average number of nearest neighbors in tellurium was found to be ~ 3 at 600°C, increasing with temperature to a value approaching 6 at 1700°C. In order to go further and convert this essentially one-dimensional information into information about the three-dimensional structure, it is necessary to make a comparison with models which are suggested by other information such as chemical bonding, valence, and known properties of chemically analogous materials. In doing

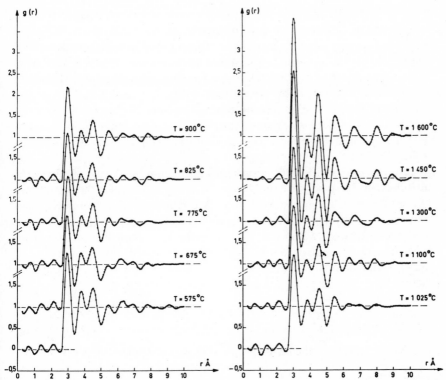

Fig. 3.10.   The quantity $g(r)$ for molten tellurium at a number of temperatures (Tourand et al., 1972).

this, Tourand *et al.* (1972) were guided by a view of the crystal structure of chalcogenide elements as a sequence arising from increasing distortion from a simple cubic structure (polonium) to a hexagonal structure (selenium). As this distortion proceeds along the diagonal axis of the cube, two neighbors come closer and four of them go farther apart, so that there is a progression from six nearest neighbors with metallic bonding to two nearest neighbors with covalent bonding. The closely-linked pairs of atoms form chains in the direction of the diagonal axis of the cube, which is now the *c* axis of a hexagonal lattice. Crystalline Te represents an intermediate stage in this sequence. To explain the structure of the liquid, Tourand *et al.* consider that a somewhat different distortion of the simple cubic structure leads to a layer-like structure with three nearest neighbors for each atom, as in the arsenic A7 structure. They find that this structure is consistent with the radii and the number of atoms in the second and third shells surrounding a Te atom. They conclude that as T is increased above 600°C, the structure gradually transforms into the more symmetrical simple cubic structure with six nearest neighbors. They suggest that as T decreases below the melting point 452°C, the coordination number decreases toward 2, and the number of two-fold and three-fold coordinated atoms are about equal at the melting point. Of course, the liquid has no long-range order, and the suggested structures are expected to exist locally about a typical atom over periods of time appreciably larger than the vibrational period. A more recent paper (Tourand, 1975) reports measurements down to 403°C which indicate that the coordination number does not decrease appreciably below 3. The implications of these results are discussed in Section 7.6.

In order to describe diffraction in a binary alloy A–B, it is necessary to introduce three partial structure factors S_{AA}, S_{AB}, and S_{BB} for the three types of pairs of atoms; each of these has a corresponding pair-distribution function g_{AA}, g_{AB}, and g_{BB}. The relations between $S_{ij}(Q)$ and $g_{ij}(r)$ are the same as the ones between S and g in Eqs. 3.10 and 3.11. The coherent scattering intensity $I(Q)$ for neutrons or X-rays depends on the atomic form factors f_A and f_B, and the concentrations c_A and c_B, as well as the partial structure factors:

$$I(Q) = c_A^2 f_A^2 (S_{AA} - 1) + c_B^2 f_B^2 (S_{BB} - 1) + 2c_A c_B f_A f_B (S_{AB} - 1)$$
$$+ c_A^2 f_A^2 + c_B^2 f_B^2. \tag{3.12}$$

For X-ray diffraction the form factors as well as the partial structure factors depend on Q.

If a single diffraction measurement is made of $I(Q)$ for a binary alloy, it is difficult to infer much from the result except in special cases, since it is a complicated average result of the three partial structure factors. For instance, it might be inferred that $I(Q)$ is or is not consistent with a particular model which was generated from essentially different information. An investigation of the structure of liquid and amorphous $Ge_{0.175}Te_{0.825}$ by neutron diffraction is an example of this approach (Nicotera et al., 1973). It is possible to derive the three different partial structure factors if $I(Q)$ is measured for the same alloy three different sets of form factors. This can be done by determining three neutron different diffraction curves or two neutron diffraction curves plus one X-ray diffraction measurement, where the alloy is changed by isotope enrichment in order to change the neutron scattering form factors (Enderby et al., 1967). If there is reason to expect the partial structure factors to be independent of concentration, it is possible to use X-ray measurements alone and measure $I(Q)$ at several concentrations (Halder and Wagner, 1967).

The only study so far of liquid semiconductors, which derives information about the partial structure factors, is work of Enderby and coworkers on Cu_2Te and CuTe (Enderby and Hawker, 1972; Hawker et al., 1973). Discussion of their results will be left to Section 8.4, but it is worth noting here some considerations about such diffraction studies on binary alloys in general. First, these studies require very special facilities (a high-neutron-flux reactor) which are to be found in only a few places in the world, and a large investment of experimental effort must be made in order to get information for even one composition. A second limitation is that the difficulty in inferring three-dimensional structural information from structure factor data, mentioned already for tellurium, is compounded when binary alloys are considered. One must proceed by setting up alternative models and determining the extent to which the experimental partial structure factors agree with those generated by the models. This problem is a common one for diffraction structure studies, and it has been overcome with brilliant success for many large biochemical molecules. The structure of a binary alloy is probably less complicated, but the difficulties of working backwards from possible models are significant. This prevents diffraction studies from being quite the straightforward means for deducing structure that one might think.

As indicated by Eq. 3.10, $S(Q) - 1$ for monatomic liquid is the Fourier component at wave vector Q of the mean fluctuation from uniform composition $g(r) - 1$. In the limit $Q = 0$, $S(Q)$ becomes simply the fluctuation from the mean density, which is related by statistical mechanical theory to the isothermal compressibility. For the case of binary alloys $A_x B_{1-x}$, the corresponding statistical mechanical relations are simpler if the density N/Ω and the atomic fraction x are used to describe the particle densities rather than the individual concentrations c_A and c_B. The three partial structure factors S_{AA}, S_{AB}, and S_{BB} are transformed into three partial structure factors S_{CC}, S_{NC}, and S_{NN} relating to concentration (C) and density (N) fluctuations by three equations involving in addition the concentrations c_A and c_B. Bhatia and Thornton (1970) have derived these relations, and have presented the relationships of $S_{CC}(0)$, $S_{NC}(0)$ and $S_{NN}(0)$ to thermodynamic parameters. The relationship of $S_{CC}(0)$ to $(\partial^2 G/\partial x^2)_T$ has been discussed already in the preceding section. We refer the reader to the original work for the other results.

CHAPTER 4

EXPERIMENTAL METHODS

The difficult experimental problems in making measurements on liquid semiconductors have undoubtedly dampened progress in this field. It seems worthwhile, therefore, to devote a chapter to a discussion of the experimental methods, with emphasis on those factors which are peculiar to semiconducting liquids. We continue to emphasize electrical measurements.

In addition to the problems related to a particular measurement, one must also contend with the following difficulties caused by the nature of liquid semiconductors:

(1) They usually have an appreciable vapor pressure of at least one constituent (Te, Se, S). These vapors (Te, Se), as well as less volatile constituents (Tl, Pb), are often poisonous. They must, therefore, be carefully controlled to avoid contaminating the laboratory or the experimenter.

(2) Liquid semiconductors are often corrosive to container or electrode materials, particularly metals, because of the high temperature and their chemical nature.

(3) The liquid is usually susceptible to oxidation.

(4) There are difficulties in maintaining a uniform, controlled composition because of volatility, the tendency to segregate on freezing, and other factors.

(5) The liquid frequently expands on freezing, and the frozen liquid may bind to the walls of the container and cause large stresses due to differential thermal contraction. As a result, there is frequently a problem due to cell breakage. The combined effects of all these difficulties lead to constraints on the experimental design which often lead to time consuming procedures or limit the experimental range or accuracy.

4.1 HIGH-TEMPERATURE TECHNIQUES

Many of the methods and elements of the apparatus needed for the study of liquid semiconductors are in the everyday domain of the metallurgist or high-temperature chemist. An excellent source of information on these techniques is provided by a monograph edited by Bockris *et al.* (1959). In addition to information and references concerning temperature measurement and the design of furnaces, it contains data relevant to container materials, thermal expansion, chemical compatibility, and stability. The descriptions of apparatus and methods for physiochemical measurements contain design concepts useful in devising methods for other types of high-temperature measurements. It is noted here also that Glazov *et al.* (1969) provide detailed descriptions of the apparatus and methods which they used in measuring the electrical conductivity, thermopower, thermal conductivity, magnetic susceptibility, density, and viscosity of liquid semiconductors.

4.2 ELECTRICAL RESISTIVITY

The resistivity ρ can be measured either by electrodeless methods, or, more directly, in cells containing electrodes. Each approach has some particular advantages and disadvantages. Electrodeless methods have the advantage that the sample can be placed in a sealed container, thus avoiding many problems due to corrosion of electrodes and difficulties in controlling the composition.

Electrodeless methods fall into two classes, depending on whether the mechanical or electrical effect of eddy currents induced in the sample is measured. In the mechanical category is the method of Roll and Motz (1957), which derives the resistivity from the torque about the axis of a cylindrical sample caused by eddy currents induced by a rotating magnetic field (Takeuchi and Endo, 1962). The controlling equation contains a term dependent on the viscosity which can be supressed by making the radius of the cylinder small. The method has been used for liquid semiconductors by Regel (1948) and by Glazov and associates (1969). Another method which does not suppress the effect of the viscous term has been used extensively by Glazov and associates (1969) in simultaneous measurements of the resistivity ρ and the kinetic viscosity η_k. (η_k is the viscosity η multiplied by the density.) In this method the liquid sample is sealed in a cylinder which is part of a torsion oscillator. The values of ρ and η_k are derived from measurements of the oscillation period and the damping rate by means of theoretical equations which relate the two pairs of parameters.

In the second category of electrodeless methods, the AC impedance of an induction coil as affected by eddy currents in the sample within the field of the coil is measured and analyzed to obtain the resistivity. This type of approach can also be used to obtain information about the Hall coefficient and the magnetoresistance (Nyberg and Burgess, 1962). In practice, only the resistivity is usually determined. This is done by analysis of the phase shift. Examples of steady state measurements are the work of Yosim *et al.* (1963) and measurements on liquid selenium by Gobrecht *et al.* (1971a). Lee and Lichter (Beer, 1972) give a detailed discussion of the use of this method for metallic alloys. Haisty (1967, 1968) has developed a transient method in which the sample is dropped through the resonance circuit coil of a radio frequency oscillator. This method requires a calibration curve for the oscillator and the particular geometry of the sample container, but it is rapid and can be used over a very wide range of resistivities (Haisty and Krebs, 1969a, b). However, the reported accuracy is limited to 10–20%.

Much more straightforward measurements are possible with the use of electrodes. The increased experimental difficulty in making and using a suitable sample cell is compensated by the possibility of obtaining much more precision, and the decreased likelihood of systematic error. This is important if the data is to be used for detailed analyses in terms of theoretical models. The precision and accuracy of results reported in the literature for electrode methods generally seem to be higher than they are for electrodeless methods. Because of spreading resistance at the current (I) leads, it is important to use a four electrode configuration, and measure the potential drop V at two electrodes located between the current electrodes, as illustrated in Fig. 4.1. For such a configuration $V = (\rho/G)I$, and the geometric factor G can be conveniently determined by calibration with a known liquid, such as mercury.

Cells constructed with sealed metal electrodes (usually tungsten) are often limited to temperatures below about 500°C (Cutler and Mallon, 1965). Although it is possible to seal metal electrodes in higher melting glasses than pyrex, most liquid semiconductors contain elements such as tellurium

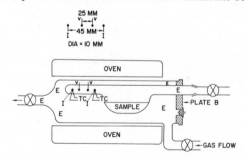

Fig. 4.1.   Glass cell for thermoelectric measurements. The oven and cell are tilted vertically when measurements are to be made (Cutler and Mallon, 1965).

or selenium which corrode the electrodes at higher temperatures. However, the use of metal electrodes at higher temperatures has been reported (Abrikosov and Chizhevskaya, 1970). In this case the electrodes were protected from corrosion by a coating of colloidal carbon.

For temperatures above 500°C, the natural materials are graphite for the electrodes and fused quartz or vycor for the container (up to $\sim 1100°C$). It is very difficult to make leak-tight seals between quartz and graphite, but a method for doing this has been described (Vukalovich *et al.*, 1967) which depends on forcing a quartz sleeve down on the graphite by means of a vacuum. Other methods which have been used include carefully-machined fits (Zhuze and Shelykh, 1965; Amboise *et al.*, 1968) and ceramic cement (Cutler and Petersen, 1970). Another possibility is to arrange the seals to be above the surface of the liquid. Dancy (1965) describes ceramic cells made with sealed graphite electrodes.

4.3 THERMOPOWER

The simultaneous measurement of the thermopower and the resistivity provides information which has been very useful in analyzing the electronic character of liquid semiconductors. The four-electrode method (Fig. 4.1) for measuring the resistivity lends itself to this approach, since one needs only to place thermocouples at two of the electrodes in order to measure the thermopower. If there is a difference in temperature at the electrodes ΔT, the potential difference ΔV for zero current is related to the thermopower S_{LM} (Bridgman, 1961) by:

$$\Delta V = S_{LM} \Delta T. \tag{4.1}$$

The quantity S_{LM} is the thermopower of the liquid in reference to the metal in the leads which are used to measure ΔV, and the value of the absolute Seebeck coefficient of the reference metal S_M is needed to calculate the Seebeck coefficient of the liquid: $S_L = S_{LM} + S_M$. The quantity S_L is referred to in theoretical formulas for the thermopower. Data on S_M for reference metals in high temperature measurements can be found in a paper by Cusack and Kendall (1958).

Although it is possible to measure the integrated thermal electromotive force of one electrode as a function of temperature with reference to a second electrode kept at a fixed temperature, and then differentiate the result to determine S as a function of T, the usual procedure is to allow both electrodes to vary in temperature, while maintaining a moderate difference ΔT between them which is small compared to average value of the absolute temperature. Although it is difficult to measure the absolute temperature with accuracy of 1°K or better, the difference in temperature between the thermocouples can be measured readily with an accuracy of ~0.1°K. But this temperature difference may differ appreciably from ΔT, which refers to the points of contact with the liquid. The accuracy of a thermopower measurement depends greatly on the requirement that the temperatures of the thermocouples reflect accurately the temperature of the electrodes. Failure in this respect is probably responsible for a good deal of error of data in the literature.

The most certain way to assure that the temperature of the thermocouple is the same as that of the electrode is to keep each electrode in a separate compartment which is at a uniform temperature. This type of approach has been used by some investigators (Bitler *et al.*, 1957; Abrikosov and Chizhevskaya, 1970), but it leads to an awkward design for the cell and apparatus. The other approach is to permit an appreciable temperature gradient in the vicinity of the electrodes, but to make sure that the junction of the thermocouple has a much lower thermal impedance to the electrode than to the rest of the system. Of course, it is important in such an arrangement that each electrode has a uniform temperature. For a sealed metal electrode, this can be accomplished by making the electrode relatively thick and short, and using very thin wires for the thermocouple which are welded to the electrodes (Cutler and Mallon, 1965). Since sealed metal electrodes are generally used at lower temperatures ($\gtrsim 500°C$) where heat transport by conduction dominates over radiation, this approach is very effective. Graphite electrodes are generally used at higher temperatures where heat transfer by radiation becomes important. In this case, the best procedure is to drill a hole in the graphite electrode and place the thermocouple junction within the tip near the surface which is in contact with the liquid (Stoneburner *et al.*, 1959; Cutler and Petersen, 1970; Dancy, 1965).

4.4 HALL EFFECT

It has been noted in Section 2.3 that the experimental methods used to measure the Hall effect in liquid semiconductors are an outgrowth of several contemporaneous studies of liquid metals which succeeded for the first time in yielding consistent results from different laboratories. A joint paper by several of these investigators discusses the special problems caused by the fluidity of the sample and the methods for dealing with them (Cusack *et al.*, 1965).

The main problems arise from the motion of the liquid as the result of the Lorenz force or thermal convection, and from the presence of bubbles. The motion of the liquid, particularly in the presence of temperature gradients, causes spurious voltages which are difficult to analyze or nullify.

The best solution is to use a very thin cell, and thus inhibit motion by means of the hydrodynamic impedance. A thin cell also has the favorable effect of increasing the magnitude of the Hall voltage, so that the main limitation to this approach would seem to be the need to fabricate a cell with a known uniform thickness. However, a thin cell also increases the problems caused by bubbles. Bubbles can cause very distressing erratic effects on the Hall signal, and their elimination is a problem affected by the cell design, filling procedure, and the chemical nature of the liquid (Cusack et al., 1965).

Liquid semiconductor measurements have been made by all of the possible electrical methods for a stationary sample, including a DC current in a DC magnetic field (Enderby and Walsh, 1966), and an AC current in a DC magnetic field (Donally and Cutler, 1968); the use of AC in both the current and magnetic field with frequency mixing for the Hall signal seems to be most popular (Perron, 1970a; Busch and Tieche, 1963a; Male, 1967).

4.5 GENERAL PROBLEMS

Aside from the technical problems in making particular measurements, there are several factors which often limit severely the mechanical efficiency of the experiment but are rarely discussed. These arise from the problems in preparing samples with a controlled composition, and the difficulties caused by the breakage of cells and excessively time-consuming procedures.

The possibility of damaging the cell when the liquid freezes in it, and the need to thoroughly mix the alloy after fusion are factors which favor the use of cells in which the liquid is introduced for measurement and then removed before cooling. Dip cells are a particularly attractive approach for doing this. The melt is prepared in a crucible (which may be expendable) and the cell with internally situated electrodes is simply dipped into the liquid. This type of cell has been used to obtain very precise resistivity data in liquid metals (Adams and Leach, 1967). Another version of this approach is to draw the liquid up into a pipette, and this seems to be amenable to measuring the thermopower in addition to the electrical resistivity.

Another approach for expediting experimental measurements is to change the composition of the alloy by adding a known amount of a constituent without cooling (and freezing) the liquid. A method for doing this without permitting air to reach the liquid or vapors to escape is illustrated in Fig. 4.2 (Field, 1967). Two side arms along the length of a tube are used to create a flowing gas baffle between the access port and the sample. An inert gas is introduced into one side arm and withdrawn at the second by means of a slight vacuum. This creates countercurrents for both the diffusion of air into the cell from the port and for diffusion of vapors from the sample out of the port, permitting safe access to the sample for changing composition.

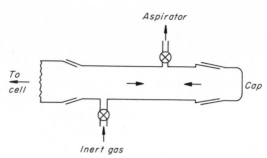

Fig. 4.2. Vapor baffle. Before the cap is removed a current of inert gas is caused to flow through the stopcocks as indicated by the arrows.

It is also possible to siphon off the liquid from the cell at the end of an experiment, and thus protect the cell from damage on cooling.

The techniques discussed above are appropriate only if the vapor pressure is relatively low, as it is in Te alloys at moderate temperatures. Even so, it is necessary to minimize distillation of the volatile constituent if part of the cell exposed to the vapor is below the condensation temperature. To minimize the rate of distillation, it is important to fill the cell with an inert gas such as argon, rather than have a vacuum. If some of the apparatus extends outside of the heated zone of a furnace, it is very desirable also to block the tube at the boundary of the hot zone with a plug of a glass or quartz wool in order to prevent convective gas transfer in this region.

Another possible technique for liquids containing a constituent with a high vapor pressure is to control the vapor pressure rather than the composition by means of a gas flow system. In that case, the composition must be ascertained from an independent study of the vapor pressure in relation to composition and temperature (Vechko *et al.*, 1967).

In the case where the vapor pressure is higher than one atmosphere, it is possible to put the cell into a chamber which is maintained at an equal or higher pressure by means of an inert gas, in order to maintain the seals in the cell. Such an approach has been used in studies of molten selenium up to a vapor pressure P of 20 atm (Gobrecht *et al.*, 1971a). The experimental difficulties become more severe as T and P are increased, but nonetheless measurements have been made above the critical point, which usually requires $P > 1000$ atm and $T > 1000°C$. These include measurements of the conductivity (Franck and Hensel, 1966), thermopower (Schmutzler and Hensel, 1972), and Hall coefficient (Even and Jortner, 1973) of mercury, and the conductivity of selenium (Hoshino *et al.*, 1976).

CHAPTER 5

THE ELECTRONIC STRUCTURE
OF LIQUID SEMICONDUCTORS

An understanding of the electronic structure is basic for interpreting the behavior of liquid semiconductors. This, and the interpretation of experimental measurements, discussed in Chapter 6, are the two major theoretical problems for liquid semiconductors. In crystalline solids, the key concept for describing the electronic structure is the energy $E(\mathbf{k})$ as a function of the wave vector $\mathbf{k}$ for the various bands. From this, various kinds of information about the electronic structure can be derived, including the density of states $N(E)$. In amorphous materials, the wave vector does not provide a good quantum number, but the energy does. Consequently $N(E)$ has been the usual starting point for interpreting the electronic structure.

Although the electronic structure is governed by the chemical composition and the arrangement of the atoms, it is common in crystalline solids to regard this as a given parameter, since the dependence of the electronic structure on the possible variation of the chemical structure is not relevant to many problems. This is not so in liquid semiconductors, since the molecular arrangements may change with temperature as well as composition. We shall be very much concerned with the dependence of electronic structure on a variable chemical structure. The dependence is mutual, but much less can be said about the dependence of the molecular arrangement on the electron structure.

Three types of models for the electronic structure can be identified:

(1) the metallic model
(2) the distorted crystal model
(3) the molecular bond model.

Each has a different first approximation or starting point, with corresponding modifications which hopefully approximate the true situation. Each model emphasizes or exaggerates some aspects of the true situation. Their successful use depends on both the extent to which the emphasized aspect corresponds to reality and whether the effects of deviations from the idealization can be successfully introduced. Liquid semiconductors encompass a wide range of behavior, and it seems likely that some of them are closer to one model or another. For instance, molten selenium is probably better represented by a molecular bond model, while highly conducting liquids such as molten tellurium or Mg–Bi alloys may be closer to the metallic model.

The use of these schemes varies between "hand waving" explanations of experimental observations, formulation of heuristic models for treating experimental or theoretical information, and exact theoretical treatments in which the approximations are well-defined. In the succeeding sections, we discuss each of these approaches. Theoretical studies frequently are confined to excessively simplified models with limited relationship to real systems, or else become so complex mathematically that they provide little physical insight. Therefore our discussion of theoretical work will be relatively brief and aimed at providing reference to some of the directions of investigation.

All three models lead to a density of states curve which contains a dip in $N(E)$ near the Fermi energy as shown by the solid curve in Fig. 5.1a. This corresponds roughly to the band gap between the valence and conduction bands in a crystalline semiconductor or semimetal. The dip in $N(E)$ is frequently referred to as a pseudogap. An important additional consideration is the spatial character of the wave functions. The states in the pseudogap may be localized rather than extended through the entire volume, and this has an important effect on their transport behavior. This aspect of the electronic structure is discussed in the last section.

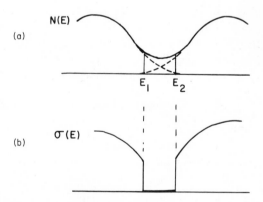

Fig. 5.1.   (a) Density of states $N(E)$ and (b) conductivity $\sigma(E)$ in the pseudogap. The states are localized between the mobility edges E_1 and E_2 (Cutler, 1974a).

5.1   THE METALLIC MODEL

The metallic model starts with the free electron approximation and then introduces the ion-electron interaction. In the case of simple metals, it has been established that the effective interaction can be represented by a weak pseudopotential, so that simple perturbation theory can be used with some accuracy (Heine, 1970). In liquid semiconductors, on the other hand, the interaction is expected to be strong.

Before considering the metallic model for semiconducting liquids, it is useful to review some aspects of its application to crystalline metals. In the absence of the crystal interaction, the unperturbed energy $E_0(\mathbf{k}) \propto k^2$. The effect of long-range order in a crystal is to eliminate all the Fourier components $V(\mathbf{k})$ of the interaction potential except at wave vectors $\mathbf{G}$ corresponding to the Bragg reflections. The resulting perturbation causes the dispersion curve $E(\mathbf{k})$ to have discontinuities at the Bragg planes as illustrated in Fig. 5.2a. As a consequence, the density of states $N(E)$ has relatively small perturbations from the (free electron) parabolic shape, which correspond to a

(a)

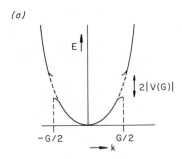

(b)

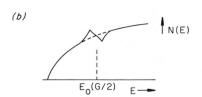

Fig. 5.2.   The ion–electron interaction in the metallic model. (a) Effect of a Bragg plane in a crystal on $E(k)$. (b) Effect of a Bragg plane in a crystal on $N(E)$. (c) Effect of a cluster interaction in a liquid on $N(E)$.

(c)

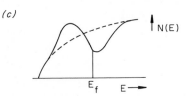

shift of states near the Bragg planes to somewhat higher or lower energies, as shown in Fig. 5.2b. If the Fermi energy E_f lies within this region, which is near $E_0(G/2) = h^2G^2/8m$, there is a net decrease in energy $\Delta E \sim |V(G)|^2$ corresponding to the change in the integrated area of $N(E)$ below E_f. The quantity ΔE is structure sensitive, that is, a distortion of the crystal at constant volume is favored if it increases the magnitude of ΔE. The deviation from the ideal c/a ratio of hexagonal close packed metals and the distortion of other simple metals from cubic symmetry have been explained by this mechanism (Heine and Weaire, 1970a).

In simple metals, $V(G)$ is usually small compared to E_f, and the resulting structure-sensitive energy terms are correspondingly small. A liquid semiconductor differs from a crystalline metal in the absence of long-range order, and in having stronger interactions. The absence of long-range order means that all Fourier components of the potential play a role, so that the sum of the perturbations in energy due to the individual Bragg planes of a crystal is replaced by an integral weighted by the structure factor $S(k)$. However, it is the short range order which is decisive in determining the structure-sensitive changes in energy. At the very least, there will be a short range order corresponding to random packing of hard spheres (Egelstaff, 1967). The quantity $S(k)$ will have strong maxima near values of $k \sim 2\pi/d_i$, where d_i are near-neighbor distances discussed in Section 3.5. The ionic potential also causes $V(k)$ to be large in magnitude in a similar range of k. Although the interactions are too large in liquid semiconductors for simple perturbation theory to be valid, the qualitative effects can be expected to be similar to those outlined for crystalline metals. A large dip may develop in $N(E)$, as illustrated in Fig. 5.2c, which is associated with a particular geometric arrangement of the atoms. If E_f is in the vicinity of the minimum, there will be a decrease in the energy which may make this arrangement, or cluster, stable with respect to alternative arrangements at the same volume.

The metallic model for the electronic structure of a liquid semiconductor was developed by Mott (1971), and it is often referred to as the pseudogap model. As discussed in detail in Chapter 6, the low value of $N(E_f)$ is directly related to the relatively small value of the electrical conductivity and to other characteristic properties of liquid semiconductors. The increased thermal energy due to increasing temperature tends to break up the clusters, with the result that the pseudogap becomes shallower, $N(E_f)$ increases, and the electrical properties revert to those typical of liquid metals.

Although the metallic model provides a simple and satisfying qualitative explanation for semiconductor behavior of liquids, the corresponding theoretical problem is extremely formidable because of the strong interactions. In terms of scattering formalism, one must take into account multiple scattering to high orders. This approach to the theory of electronic structure has been investigated by Edwards (1962), but progress in this direction has not been sufficient to shed much light so far on the behavior of real systems.

We have not considered the effect of having more than one kind of atom. Most liquid semiconductors are alloys, often with components which differ appreciably in electronegativity. The existence of different kinds of atoms in a metal adds an important dimension, and the theory of binary metallic alloys has long been an important problem. In recent years, important progress has been made by the use of mean field theories such as the Coherent Potential Approximation (Soven, 1967, 1969; Velicky *et al.*, 1968). In this class of models an effective field is postulated which represents the average environment for each atom, and it is arrived at by a self-consistent method. These methods are amenable to model calculations which yield the density of states in disordered systems, so that the behavior of $N(E)$ can be determined at some level of approximation as a function of the composition of an alloy. The basic concepts of the Coherent Potential Approximation are not limited to metals, and there have been applications to semiconductors or insulators (Sen and Cohen, 1972).

5.2 DISTORTED CRYSTAL MODEL

Another idealization for the electronic structure of disordered materials is the crystalline phase of a semiconductor. This provides a starting point where a band structure with a band gap is already present. The nature of the "disorder" as applied to a crystal is often not clearly defined, and depends on the nature of the disordered material. Consequently, the implications derived from this type of model may be somewhat vague.

Two effects of the disorder can be identified. Topological disorder is the loss of long-range order while some kind of short-range order persists. It may be expected to blur the crystal potential in a way which is analogous to line broadening due to small crystal size in diffraction theory. When this is translated to effects on the electronic structure, the irregularities (due to van Hove singularities) in the crystalline density-of-states curve are smeared as shown in Fig. 5.3a and b. A second effect is specifically due to fluctuations in atomic density and potential. This leads to tails on the band edges, which are discussed in Section 5.4. In principle these tails extend across the gap, so that the gap becomes a pseudogap as shown by the dashed curves in Fig. 5.1a.

A systematic theoretical development of the disordered crystal model has been carried out by Gubanov (1965). He introduces perturbation parameters which describe deformations from a perfect crystal, and derives equations describe the corresponding effects on the wave functions and energies of the resulting states. In the limit where the perturbation is small, the system becomes a crystal.

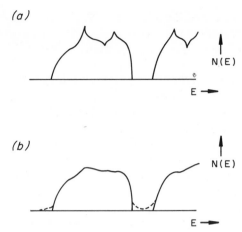

Fig. 5.3. Distorted crystal model. (a) $N(E)$ with van Hove singularities in a crystal.
(b) Smoothed $N(E)$ in the distorted crystal. The dashed lines indicate the effect of potential
fluctuations.

5.3 THE MOLECULAR BOND MODEL

Another idealized state for liquid semiconductors consists of discrete molecules which are far enough apart so that the electronic levels are discrete. In
contrast to the other two models, one starts with gaps but no bands. When
molecules are brought together, the discrete electronic energies broaden into
bands appropriately described by tight-binding theory. As in the distorted
crystal model, band tailing is expected to cause overlap between the bands
to form a pseudogap, as indicated by the dashed lines in Fig. 5.1a.

For discrete molecules, it is often a reasonable approximation to assume
that the electronic states of different bonds and nonbonding electrons are
independent. The perturbation to these bond states caused by bringing molecules close together in a liquid semiconductor may be comparable to that
due to the interaction between the bonds within a molecule. Therefore it
seems reasonable to use the individual bonds as the starting point for liquid
semiconductors. We refer to this as the bond orbital model. The bond orbital
model has the added convenience that it applies equally well when bonding
leads to network structures, as in As_2Se_3, or when small molecules are the
result as in Tl_2Te.

An important advantage of the bond orbital model is that the semi-
empirical methods of structural chemistry can be used for making estimates
of the energy and electronic character of the bond orbitals, and from this
the electronic bands can be inferred by the tight-binding model. The bond
orbital model was first applied to liquid semiconductors in the case of Tl–Te
alloys (Cutler, 1971a, b), and this example serves well to describe the method.

Consider first covalent bonding of Te atoms. The occupied orbitals of Te atoms are $(5s)^2(5p)^4$. The bond orbital is assumed to be a linear combination of 5p orbitals of the two atoms which lie in the direction of the bond (say p_z). As the atoms are brought together, the energy decreases for the symmetric linear combination of atomic wave functions which concentrates charge between the atoms. Conversely, the antisymmetric combination which cancels the charge density between the atoms increases in energy, causing a splitting shown in Fig. 5.4a. The lower energy state is commonly called the σ(bonding) state, and the other is referred to as a σ^* (antibonding) state. A tellurium atom normally forms two bonds, the second (say the p_y orbital) at right angles to the first. The four valence p electrons of the Te atom go into the lowest energy states, one in each bond (which it shares with another atom), and two in the remaining p_x nonbonding (π) orbital. This yields a net decrease in energy which is the binding energy. If the tellurium atom were to form a third bond with the p_x orbital, only one of the electrons in the original nonbonding orbital could occupy the lower energy σ state, and the other would have to go into the higher σ^* state. Since the splitting is symmetrical in the first approximation, there is no reduction in energy and a third bond is normally unstable. (Circumstances under which three-fold bonding may be stable for Te are discussed in Section 7.6.) The right-hand side of Fig. 5.4a indicates the broadening into bands, which results from the tight-binding interaction. Each band contains $2N$ states, where N is the number of tellurium atoms.

In all of this, we have ignored hybridization of the atomic orbitals. Some mixing of atomic orbitals from higher or lower orbitals of a given atom generally yields an increase in the final bond energy. This is accompanied by a change from the 90° bond angle for the pure p-bonds just described. In simple molecules the measured bond angle can often be used to infer the nature of the mixing (Pauling, 1960).

The band structure arrived at in this way for tellurium can be identified with the electronic structure of crystalline tellurium derived from conventional band calculations. Indeed, the LCAO method, which is the quantitative expression of the molecular orbital approach, is one of the methods which has been used in band calculations for tellurium and selenium (Reitz, 1957). Figure 5.5 shows the band structure of tellurium calculated by Treusch and Sandrock (1966). The three bands corresponding to the bonding σ orbitals, the nonbonding π orbitals, and the antibonding σ^* orbitals can be readily identified. In the crystal, the tellurium atoms are arranged as spiral chains parallel to the c axis. The nearest neighbor distance r_1 for atoms within a chain is significantly shorter than the distance r_2 between adjoining atoms on different chains. The ΓZ direction is along the chain axis, and the dispersion in energy in this direction indicates roughly the effects of interactions between bond orbitals within the chain. In the K direction there is a pronounced dispersion which can identified with the interactions between the

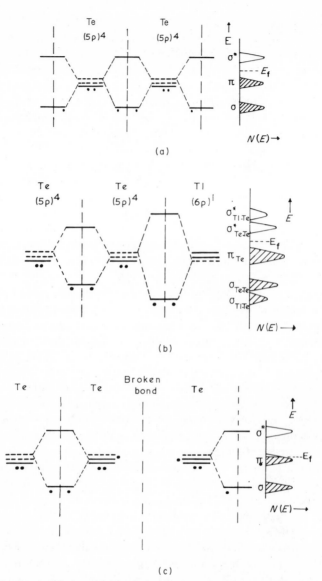

(a)

(b)

(c)

Fig. 5.4. Bond orbital model. The dashed levels are atomic orbitals which are converted to bond orbitals with levels indicated by solid lines. The dots represent electrons in the ground states. The vertical dashed lines separate domains of each atom. The tight-binding bands generated by these levels are shown on the right. (a) Te–Te bonds. (b) Energy level scheme postulated for Tl–Te and Te–Te bonds. (c) Effect of breaking a Te–Te bond. Each dangling bond adds two electron states and one electron to the π band, giving rise to one hold (Cutler, 1971a).

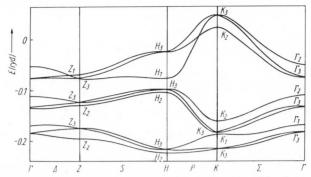

Fig. 5.5.   Band structure of crystalline tellurium (Treusch and Sandrock, 1966).

chains. The band structure for Se (Treusch and Sandrock, 1966) is very similar to that for Te. The dispersion in the K direction is considerably smaller for Se, which reflects the smaller intermolecular interaction as implied by the larger ratio r_2/r_1 in the selenium crystal.

Turning now to Tl–Te alloys, a bond formed between Te and Tl would use the 6p atomic orbital of Tl. The splitting into σ^* and σ bond orbitals would be different than for Te–Te, reflecting a different bond energy. The electronic structure of a molecular system containing Tl–Te and Te–Te bonds is shown in Fig. 5.4b, and the resulting density of states curves are indicated schematically on the right.

An important feature of this scheme is that the number of states in each tight-binding band is determined by the number of bonds and atoms. Since this information provides also the number of electrons, the scheme also fixes the position of the Fermi energy E_f at $T = 0$. In the case of pure Te or Tl_2Te, E_f is above the nonbonding π band due to the tellurium atoms. It is clear that any completely bonded covalent molecular system containing chalcogenide elements would have the Fermi energy above the bands formed from nonbonding electrons of the chalcogenide atoms.

The bond orbital model also provides some insight into possible effects of temperature on the electronic structure. Thermal vibrations which distort the configuration of covalently bonded atoms will reverse the splitting process and move the energies of the discrete states back toward their original values corresponding to the atomic orbitals. Thus thermal vibrations would cause the σ^* and σ to have tails extending toward the center, as shown in Fig. 5.6. Another possibility is that some bonds will break, causing the bonding and antibonding states to move back into the nonbonding band. When this happens to a Te–Te bond, there will be two extra states and one extra electron in the valence band for each dangling bond, so that the net effect is to add a hole to the band, as shown in Fig. 5.4c. This mechanism has

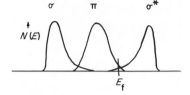

Fig. 5.6. Thermal distortion of the σ and σ^* bands due to bond stretching (Cutler, 1971a).

been proposed by the author (Cutler, 1971a, b) in order to explain the S-type behavior of Te–rich alloys of Tl and Te, and its application is discussed in detail in Section 7.3. There are reasons for expecting that broken bonds may provide the more important mechanism for thermal change in the electronic structure in the case of chain-molecules, whereas distorted bonds may be dominant in network structures (Cutler, 1971a).

The decrease in the intermolecular interaction relative to the bonding interaction, as exemplified by the greater value of r_2/r_1 for selenium than tellurium, occurs even more strongly for sulfur. This reflects a general difference between covalent bonding of heavy and light elements. For light elements, which provide the more familiar examples of covalent bonding, intermolecular bonding is weaker and intramolecular bonding is stronger (Phillips, 1973). The reason is related to the fact that heavy elements have a large ion core, and the higher atomic orbitals are not as well separated in energy as in light atoms. As a result, atomic orbitals other than those of the valence electrons play a greater role in bonding and band formation than what is suggested by the simple bond orbital and tight-binding scheme outlined above. Besides causing secondary bonding, the participation of these nearby orbitals causes the covalent bonds in heavy elements to be less directional, particularly in a condensed phase. Thus covalent bonds in heavy element alloys can be expected not only to be weaker with respect to configurations typical of metallic alloys, but also less resistant to distortions which change the bond angle. These strong secondary interactions can be considered to add a partial metallic character to covalent bonding in heavy metal compounds.

Many of the theoretical approaches for considering the electronic structure of disordered materials seem to be related to the bond orbital model. Perhaps the one most directly related is the tight-binding model used by Weaire and Thorpe (1971), which has shed light on some aspects of the pseudogap. This model assumes that short-range order characteristic of covalent bonding exists, but interactions occur at most between nearest neighbor atoms, so that long-range order is not a factor. This is seen in the expression for the model Hamiltonian:

$$H = \sum_{a,b \neq b'} V_1 |\phi_{ab}\rangle\langle\phi_{ab'}| + \sum_{a \neq a',b} V_2 |\phi_{ab}\rangle\langle\phi_{a'b}|. \tag{5.1}$$

The quantity ϕ_{ab} is a wave function associated with atomic site a and bond b. Thus V_1 determines the interaction between different bonds on the same atom, and V_2 determines the interaction between neighboring atoms which enter into a bond. Information is derived about the density of states by use of matrix methods. The most important conclusion from this work has been the general one that topological disorder (i.e., the lack of long-range order) does not in itself remove the band gap. It seems possible to extend this approach to include longer-range interactions, such as next-nearest neighbor interactions, and to consider systems containing more than one kind of atom. It remains to be seen how fruitful such efforts will be in the face of the greater mathematical complexity.

We mention two other theoretical approaches for disordered systems which do not fall neatly in the bond orbital or other categories of models. One of them centers on a cluster of N atoms with a specified configuration. The electronic interactions within the cluster are calculated accurately, but the effects of interactions with atoms outside of the cluster are introduced by an approximate method (Keller and Ziman, 1972; Keller and Fritz, 1974). Another method considers crystals with unit cells with a large enough number N of atoms to approximate a region of an amorphous solid, and conventional band theoretical calculations are made (Joannopoulis and Cohen, 1973). Both of these methods yield more representative results as N N is made larger. Since the computing effort goes up very rapidly with increasing N, these approaches are limited by computer economics. The reader is referred to a review paper by Thorpe and Weaire (1973) on this subject.

5.4   BAND TAILING AND ANDERSON LOCALIZATION

It was recognized early in the study of crystalline materials that spatial fluctuations in the potential due to impurities in a semiconductor cause the density of states to form a tail below the normal band edge. This is very evident if one uses a particle-in-a-box model for the states near the bottom of the band as illustrated in Fig. 5.7, where now the potential has functuations instead of being flat, There has been considerable study of this problem mainly in the context of impurity bands in highly doped semiconductors. The theory which has been developed seems to be largely applicable to amorphous materials, in view of the evidence, discussed in the preceding section, that long-range disorder does not in itself change the character of the band edge from that found in crystalline material. The theory for band tailing derived by Halperin and Lax (1966, 1967) is frequently referred to. They derive a density of states curve appropriate for the low energy tail which is of the form of $\exp[-|E|^n]$, where n may vary with E in the range $\frac{1}{2}$ to 2.

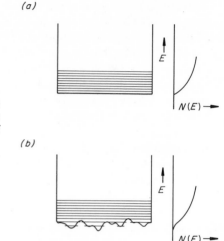

(a)

Fig. 5.7. Band tailing due to fluctuations. (a) $V(x)$ and $N(E)$ for the particle-in-a-box model with a smooth potential. (b) Effect of fluctuations in $V(x)$ on $N(E)$.

(b)

Let us consider the character of electronic states at energies in or near the band tail. Within the band, the states are extended, that is, the wave functions occupy the entire volume. This may not be so in the band tail. These states may be localized in the sense that an electron placed in one region with a negative potential fluctuation will not diffuse at zero temperature to other regions with corresponding potential fluctuations. Mott (1968) has argued that truly localized states do occur, but this has been disputed, and the question is still unsettled (Thouless, 1972). Taking the frequently accepted view that localized states do occur in the presence of fluctuations, one may ask whether the localized and extended states are separated in energy. Mott (1968) has broadened a theoretical result derived by Anderson (1958) to show that there should be a sharp boundary between the energy ranges of extended and localized states; this theory of "Anderson localization" is described below.

Anderson's theory was developed in the context of spin states, and considers whether weakly overlapping atomic states form extended states. In the absence of fluctuations tight-binding theory shows that extended states occur in a band of width $\Gamma = 2Jz$, where J is the overlap integral and z is the coordination number. The problem becomes much more difficult if there are fluctuations ΔV in the potential, since one must consider the effects of overlap between wave functions which have a distribution of different energies. Anderson's work showed that nonlocalized states would exist provided that the ratio $\eta = \Gamma/\Delta V$ exceeds a certain value η_c. For the case of $z = 6$, he estimated that $\eta_c \cong 5$. More recent studies yield a value closer to 2 (Mott, 1972a).

Mott (1968) extended Anderson's theory to consider whether localization occurs at different energies in the band. He argued that the determining factor is the density of states $N(E)$. If $N(E)$ is large enough, an electron initially situated in a particular region will always find sites at an energy close enough to tunnel to (i.e., the overlap integral is large enough) and so diffuse throughout the volume. But since the tunneling distance required to find another state at a given energy varies as $[N(E)]^{-1/3}$, there is a critical density of states $N_c(E)$ for a given ΔV below which the average tunneling distance becomes too large. The electron becomes trapped in a given region and thus has a localized wave function. This argument can be related to classical mathematical percolation theory. According to this theory, if points are connected in space by a network of linkages with varying probabilities ξ of interchange for a mobile particle, then there is a critical value ξ_c for the average probability $\bar{\xi}$ which determines whether the particle has a finite probability of being anywhere in the network, or a zero probability except within a small region. The value of ξ_c depends on the coordination number of the network and its dimensionality. Mott's argument is simply that $\bar{\xi}$ is a continuous function of $N(E)$, which in turn is a continuous function of the energy. So there is a sharp boundary at an energy E_c which separate states with $\bar{\xi} < \xi_c$, which are localized. This result is indicated in Fig. 5.1a, where E_1 and E_2 are the values of E_c for the two bands. In the following chapter we discuss the theory for transport in states at energies near E_c. However it is noted here that the electrical conductivity at finite temperatures is expected to drop abruptly several orders of magnitude for the localized states, as shown in Fig. 5.1b, so that E_c is commonly referred to as the mobility edge, in analogy to the band edge.

CHAPTER 6

INTERPRETATION OF
MEASUREMENTS

In earlier chapters, we have reviewed experimental information about semiconducting liquids, but have made only limited and generally empirical inferences from the data. The proper interpretation of experimental data, particularly in reference to the electronic structure, constitutes a major theoretical problem for liquid semiconductors which is considered in this chapter. It is natural to extend or adapt to liquid semiconductors the theories used successfully for other types of materials, such as crystalline metals and semiconductors. But it is necessary to reexamine the basic assumptions and reevaluate the approximations which are made. One difference already noted for liquids is the absence of long-range order. This means that the wave-vector does not provide a good quantum number, although it may be defined if scattering is not too strong. A second difference is that electron scattering is usually very strong, so that the Boltzmann transport equation is not valid. As a result, a different starting point must be used for the theory of electronic transport. The basis for interpreting transport data is still in a state of development and important discrepancies exist—in the Hall effect, for instance. Also, for some parameters such as the electrical conductivity and the thermopower, where there are some reasons to believe that the present theory is correct, there are only a few cases in which detailed agreement has been demonstrated between theory and experiment. Consequently, it is not certain that all pertinent aspects of behavior are adequately covered by the theory.

Outside of electronic transport measurements the situation is more obscure, largely because of the paucity of experimental studies. Since it seems likely that some of these measurements may ultimately provide important means for gaining insight into semiconducting liquids, it is desirable to review the present understanding about these phenomena. Therefore, the discussion in this chapter will encompass, in addition to the major types of electronic transport measurements, other potentially important measurements including optical and magnetic measurements.

6.1 THERMOELECTRIC TRANSPORT

6.1.1 General Equations

It has already been noted that electronic states in disordered systems are best characterized by their energy. Consequently, it is useful to discuss the electrical conductivity σ in terms of $\sigma(E)$, the two being related by:

$$\sigma = -\int_{-\infty}^{\infty} \sigma(E)(\partial f/\partial E)\, dE, \qquad (6.1)$$

where f is the Fermi–Dirac distribution function:

$$f = [1 + \exp\{E - E_f\}/kT\}]^{-1}. \qquad (6.2)$$

This expression assumes only that σ can be expressed as the sum of processes which occur at different energies, and it is very general. If, in addition, the motion of an electronic charge $-e$ at an energy E causes a transfer of an amount of energy equal to E, then, as shown in Appendix A, the thermopower S is given by:

$$S = \frac{k}{e}\int_{-\infty}^{\infty} \frac{\sigma(E)}{\sigma}\left[\frac{E - E_f}{kT}\right]\frac{\partial f}{\partial E}\, dE. \qquad (6.3)$$

These expressions are expected to be valid in most situations encountered in liquid semiconductors. If the motion of charge causes an energy transfer (positive or negative) in addition to the energy of the moving electron, the expression for S is more complicated (Fritzsche, 1971). We know of no evidence that nonelectronic heat of transport plays a role in liquid semi-conductors, and we shall generally assume that Eq. 6.3 is valid.

The physical implications of Eqs. 6.1 and 6.3 can be clarified if it is recognized that $-\partial f/\partial E$ is a bell-shaped function of width $\cong kT$, centered on E_f, which decreases as $\exp(-|E - E_f|/kT)$ when $|E - E_f| \gg kT$. Also the factor $(\partial f/\partial E)(E - E_f)/kT$ in Eq. 6.3 is very similar in shape and character to $-\partial^2 f/\partial E^2$. The functions are illustrated in Fig. 6.1a. If $\sigma(E)$ changes relatively slowly near E_f on the energy scale of kT, these two factors act like a delta function and its derivative respectively, and one gets the well-known "metallic" approximation, illustrated in Fig 6.1b:

$$\sigma \cong \sigma(E_f), \qquad (6.4a)$$

$$S \cong -(\pi^2 k^2 T/3e)(\partial \ln \sigma/\partial E)_{E_f}. \qquad (6.4b)$$

Another approximation is useful when E_f is in a mobility gap where $\sigma(E)$ is zero or negligible. If there is an abrupt mobility edge at energy E_1, as illustrated in Fig 6.1c, the Maxwell–Boltzmann (MB) approximation is

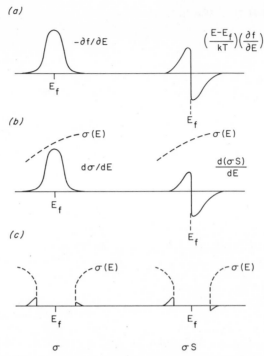

Fig. 6.1. Genesis of σ and σS from $\sigma(E)$. (a) Shape of $-\partial f/\partial E$ and $[(E - E_f)/kT](\partial f/\partial E)$. The dashed curves show $\sigma(E)$ in (b) the metallic situation and (c) the Maxwell–Boltzmann situation, and the solid curves show the integrands of σ and σS as a function of E.

accurate for $E_1 - E_f \gtrsim 4kT$, and the result is (Cutler and Mott, 1969):

$$\sigma = \sigma(E_1) \exp[-(E_1 - E_f)/kT], \qquad (6.5a)$$

$$S = \frac{k}{e}[1 + (E_1 - E_f)/kT]. \qquad (6.5b)$$

For transport by holes, the corresponding approximation is obtained by reversing the signs of e and $E_1 - E_f$ in these equations. The same approximation can be used for transport by hopping in localized states in a small range of energy at E_1 when $|E_1 - E_f| > 4kT$.

Much of the existing transport data for liquid semiconductors is in a range ($50 \gtrsim \sigma \gtrsim 500$ ohm^{-1} cm^{-1}) where the distance of E_f from the band edge is comparable to kT. In these cases, the use of the metallic or Maxwell–Boltzmann approximation can lead to serious errors. One should carry out the integrals in Eqs. 6.1 and 6.3. Practical methods for doing this are discussed in Appendix B.

6.1.2 Weak Scattering

Let us now consider the form of $\sigma(E)$ in different domains of behavior, starting with weak scattering. The function $\sigma(E)$ can be expressed in a general form in terms of the density states $N(E)$:

$$\sigma(E) = e^2 D(E)N(E), \tag{6.6}$$

where D is a diffusion constant. In the case of weak scattering, which occurs in liquid metals, the electrons have a gas-like motion with a velocity $v(E)$ and a scattering time $\tau(E)$. In this case, the Boltzmann transport equation is valid and it yields:

$$D_B(E) = v^2 \tau / 3\pi^2. \tag{6.7}$$

The well-known Ziman expression (Ziman, 1961, 1967) for τ is based on the Born approximation for electrons in single electron states which are scattered by a weak pseudopotential $u(k)$:

$$\tau^{-1} = \int_0^1 4|u(k)|^2 S(k)(k/2k_f)^3 \, d(k/2k_f), \tag{6.8}$$

where the subscript f indicates a quantity measured at the Fermi energy. This expression already contains the metallic approximation. The function $S(k)$ is the structure factor for the liquid, discussed in Section 3.5, which can be obtained experimentally from diffraction measurements. This formula has also been used successfully with $S(k)$ provided by the Percus–Yevick theory for a liquid, which is based on a hard sphere model (Ashcroft and Lekner, 1966). There are some difficulties in applying the Ziman formula in a definitive way because there is no unique pseudopotential. However its successful use in a number of problems, including the case of liquid mercury, where there had seemed at first to be a discrepancy, seems to indicate that it provides, at least in principle, a very good description of transport in simple liquid metals. An extensive discussion of weak scattering theory is given by Faber (1972).

Behavior characteristic of liquid semiconductors is increasingly found as σ falls below ~ 3000 ohm^{-1} cm^{-1}. Approximate calculations indicate that the scattering distance $\lambda_s = v_f \tau$ becomes comparable to the deBroglie wavelength ($\lambda_s k_f \sim 1$) when $\sigma < 3000$ ohm^{-1} cm^{-1}. In this range, it is clear that the weak scattering model is not appropriate.

6.1.3. Diffusive Transport

For transport with a short mean free path, it is necessary to use a starting point other than the Boltzmann equation for defining the transport parameters, such as the Kubo–Greenwood formula (Greenwood, 1958) or the

Luttinger equations (Luttinger, 1964). These formulations lead to an expression for the diffusion constant in Eq. 6.6 in terms of a velocity correlation function:

$$D_K(E) = \tfrac{1}{3} \int_0^\infty \langle v(0)v(\tau)\rangle_c \, d\tau. \tag{6.9}$$

The quantity τ is time and the subscript c indicates an average over all configurations of a canonical ensemble representing the system.

For the limit of very strong scattering, Mott (1969) has introduced a simple model in which the electron wave function is regarded as a linear combination of orbitals centered on the atoms, with a completely random phase relation between the orbitals. This leads to diffusive motion of charge in an electric field with a path length equal to the interatomic distance a. More complete derivations have been given by Hindley (1970) and by Friedman (1971). Hindley's result can be stated as follows:

$$D_K = (2\pi\hbar^3 a\lambda^2/3m^2)N(E). \tag{6.10}$$

The quantity λ is a dimensionless parameter which describes the momentum matrix element $\langle n'|p|n\rangle$ between adjoining pairs of atomic orbitals:

$$\lambda^2 = \left\langle \sum_{n'} |\langle n'|p|n\rangle|^2 \right\rangle_c (a^2/\hbar^2). \tag{6.11}$$

Friedman's theory uses a quasi-lattice formulation (Bloch tight-binding theory) with a coordination number z and an overlap integral J. It is related to Hindley's result by:

$$\lambda\hbar^2/ma^2 = z^{1/2}J, \tag{6.12}$$

where J is related to the bandwidth Γ by:

$$\Gamma = 2zJ. \tag{6.13}$$

Putting together Eqs. 6.6 and 6.10, the result is:

$$\sigma(E) = A[N(E)]^2, \tag{6.14a}$$

where:

$$A = 2\pi\hbar^3 e^2 \lambda^2 a/3m^2. \tag{6.14b}$$

This expression is of major importance for liquid semiconductors because a large fraction of experimental data for σ and S lie in a range where diffusive transport occurs. The physical meaning of this result can be seen more clearly if we express D_K in Eq. 6.10 as $a^2 v_e/6$, where v_e is the frequency of motion between sites. Then Eq. 6.10 can be rewritten with the help of Eqs. 6.12 and 6.13 in the form:

$$v_e/v_b = (2\pi^2/z)N(E)/N_b, \tag{6.15}$$

where v_b is the "band-width frequency" Γ/h and N_b is the "band-width density of states" $(a^3\Gamma)^{-1}$. This indicates that $D_K(E)$ becomes smaller in proportion to $N(E)$ because an electron at a given site has a smaller probability of finding an adjoining site at the same energy.

In his discussions of diffusive transport and the conductivity at the mobility edge (considered in the next subsection), Mott (1969, 1971) has frequently expressed his results in terms of $g = N(E)/N^0(E)$, where $N^0(E)$ is the density of states in the free-electron approximation; many other authors have followed this practice. This formulation is derived from the pseudogap model discussed in Section 5.1. We think that it sometimes causes an unnecessary ambiguity since it is not always clear which electrons should be counted in determining $N^0(E)$. (This problem is discussed in reference to Tl–Te alloys in Section 7.2.) Therefore we express the results directly in terms of $N(E)$. This has the advantage of putting the transport equations in a form which is not dependent on the validity of the pseudogap model.

6.1.4 The Mobility Edge

As discussed in Section 5.4, potential fluctuations cause states to be localized for energies lying below a critical value E_c, which is called the mobility edge. It will be seen later that the value of $\sigma(E_c)$ plays an important role in analysis of the behavior of liquid semiconductors. Mott (1970) has estimated $\sigma(E_c)$ by assuming that $N(E)$ in the vicinity of E_c is determined primarily by the mean fluctuation ΔV through the relation:

$$N(E) \cong (a^3 \, \Delta V)^{-1}. \tag{6.16}$$

When this is inserted into Eq. 6.14 and λ^2 is expressed in terms of the band width by Eqs. 6.12 and 6.13, an expression is obtained for $\sigma(E_c)$ which contains a factor $(\Gamma/\Delta V)^2$. This is set equal to η_c^{-2}, where η_c is the critical ratio of Anderson's theory. Using more current estimates for η_c and some refinements which need not be discussed here, Mott (1972a) has derived a value:

$$\sigma(E_c) = 0.025e^2/\hbar a = (610/a^*) \quad \text{ohm}^{-1} \, \text{cm}^{-1}, \tag{6.17}$$

where a^* is the interatomic distance in angstrom units. The function $\sigma(E_c)$ is approximately proportional to z^{-1} and the constant in Eq. 6.17 corresponds to $z = 6$.

Since a is typically ~ 4 Å in liquids, Eq. 6.17 suggests that $\sigma(E_c) \sim 150$ ohm^{-1} cm^{-1} provides a lower limit for the diffusive range. Mott has suggested that the upper limit be taken to be $\sigma \sim 3000$ ohm^{-1} cm^{-1} which is where weak-scattering theory breaks down.

Cohen and co-workers (Eggarter and Cohen, 1970; Tauc, 1974) have suggested that $\sigma(E)$ is also modified by an excluded volume effect at $E \gtrsim E_c$. Their argument is that the electron wave function will be completely excluded from an appreciable fraction of the volume (or severely attenuated) as E approaches E_c because of repulsive potential fluctuations. As a result, σ decreases more rapidly than $[N(E)]^2$ as $E \to E_c$, and they calculate the correction factor by means of a semiclassical model for heterogeneous transport. This model incorporates percolation theory, and leads to the conclusion that $\sigma(E)$ goes to zero at E_c continuously rather than discontinuously as suggested by Mott (1974a). On the basis of the limited existing information about mobility edges in liquid semiconductors, discussed in Chapters 7 and 8, it seems likely that the disputed differences in the shape of $\sigma(E)$ near E_c are unimportant for liquids because they occur on an energy scale $\lesssim kT$. We shall use Eq. 6.14a, together with the assumption that $\sigma = 0$ for $E < E_c$, as a practical model for the mobility edge.

6.1.5 Hopping

The states below the mobility edge are localized. The motion of charge is still diffusive, but now the electronic motion occurs on a time scale which is governed by the vibrational motion of the atoms. As the atoms vibrate, the electronic configuration changes in conjunction with the atomic configuration, and at the end of a vibrational period the electron may have moved to another position. Thus the upper limit of the frequency factor in the diffusion constant for hopping D_H is the vibration frequency $v_v \sim 10^{13}$ sec^{-1}, rather than the electronic frequency $v_e \sim 10^{15}$ sec^{-1} which occurs in Eq. 6.15. Because of fluctuations in potential, the motion of a charge to another site of equal energy is generally impeded by a barrier W, which has the effect of reducing the frequency of hopping by the Boltzmann factor $\exp(-W/kT)$. Thus one can write (Mott, 1969):

$$D_H = [a^2 \phi v_v/3] \exp[-W(E)/kT],   \qquad (6.18a)$$

where:

$$\phi = (R/a)^2 \exp(-2\alpha R).   \qquad (6.18b)$$

The factor ϕ makes a correction to the diffusion constant which allows for the possibility that the distance between sites has a larger value R instead of the interatomic distance a, and the exponential factor expresses the smaller probability for tunneling between sites when R becomes larger than the attenuation distance α^{-1} of the wave function of the electron. As E decreases below E_c, W is expected to increase and ϕ is expected to decrease.

The proper theoretical treatment for hopping was first developed by Miller and Abrahams (1960), and more recently there has been considerable study of the subject. It is a complicated problem involving the motion of charge through a network of sites which are random in energy and distance. We refer the reader to a paper by Butcher (1974) for a theoretical discussion in which the results are expressed in terms similar to those used here (Eqs. 6.1, 6.3, and 6.6). At higher temperatures, hopping between nearest neighbor sites will dominate, and W in Eq. 6.18a represents an average barrier energy for paths between pairs of sites at the same energy. Other paths exist which involve pairs of sites which are farther separated, whose impedance is larger because of the tunneling factor ϕ in Eq. 6.18b. The impedance varies as $\exp[-(W/kT) - 2\alpha R]$, and Mott has pointed out that at lower temperatures, paths with larger R will be favored since they are more likely to have a smaller W (Mott, 1969). This accounts for the decrease in activation energy of σ which is observed in many substances as T is decreased. Mott derived an expression $\sigma \propto \exp[-B/T^{1/4}]$ for this phenomenon, where B is a constant, and this is in agreement with the observed behavior in a number of disordered solids. However, it seems likely that the diffusive motion of atoms in a liquid, which often occurs on a time scale $\gtrsim 10^{-9}$ sec, may provide a lower impedance transport mechanism so that extended range hopping may not be observed in liquids (Mott, 1971).

6.1.6   Characteristic Ranges of σ

We have outlined the currently accepted theory for $\sigma(E)$ and have found that three domains are defined. For $\sigma(E) \gtrsim 10^4$ ohm^{-1} cm^{-1}, there is gas-like electronic motion with a long mean free path. The Ziman theory applies here, and this theory seems to be well supported by experiment. The second domain, where diffusive motion by nonlocalized electrons occurs, is expected to be in the range $200 \gtrsim \sigma(E) \gtrsim 3000$ ohm^{-1} cm^{-1}. The most direct experimental support for this theory is obtained from NMR studies by Warren (1971), and from the correlation of the behavior of σ with the Knight shift or magnetic susceptibility, as discussed in Section 6.4. Other experimental support for this theory in liquids consists in successful use of the theory in analyzing transport data in Tl–Te alloys (Cutler, 1974a, 1976b), which will be discussed in Chapter 7. The third domain, where $E < E_c$ and transport is by hopping, has its experimental confirmation mainly in amorphous solids.

It is important to note that because of the abrupt decrease in $\sigma(E)$ at mobility edge, with $\sigma(E) \gtrsim 1$ ohm^{-1} cm^{-1} for $E < E_c$, electronic transport in the range $1 < \sigma < 200$ ohm^{-1} cm^{-1} is expected to be caused by electrons excited thermally to extended states above the mobility edge. There has been relatively little study of liquid semiconductors in the conductivity range $\sigma < 1$ ohm^{-1} cm^{-1}, so that there is little experimental basis for judging for liquids the applicability of the hopping theory outlined above. In a recent study of the frequency dependence of σ in Se–Te alloys, however, Andreev (1973) comes to the conclusion that hopping is important in liquid semi-conductors only when $\sigma < 0.1$ ohm^{-1} cm^{-1}.

In the later chapters, we consider in detail some of the inferences which can be made from the experimental data for σ and S in terms of the theories described here. Some general observations can be made, however. Much of the data for liquid semiconductors is in the diffusive range: $200 < \sigma < 3000$ ohm^{-1} cm^{-1}. Since E_f is expected to be near or above the mobility edge for these liquids, it should be within the valence or conduction band. If, in addition, σ increases rapidly with T, as it does in many Te-rich alloys, an explanation should be sought in terms of an increase in $\sigma(E_f)$, rather than in terms of excitation of carriers across an energy gap, as is customary in semi-conductor theory.

6.2   THE HALL EFFECT

As noted in Section 2.3, a number of experimental observations indicate that the theory for the Hall coefficient R_H based on weak scattering is accurate for a number of pure liquid metals, but there is considerable question about the theory for liquids with lower electrical conductivities. Several theoretical discussions have suggested that R_H will be smaller than $1/ne$ in metals with strong scattering (Ziman, 1967; Fukuyama et al., 1969). The diffusive trans-port mechanism is fairly well established for nonlocalized electrons in the domain $\sigma \gtrsim 3000$ ohm^{-1} cm^{-1}. It is therefore appropriate to discuss Fried-man's theory for the Hall effect, which is based on the random phase model (Friedman, 1971).

In an isotropic system such as a liquid, the conductivity tensor σ_{ij} is diagonal in the absence of a magnetic field, and for the random phase model σ_{ii} is given by Eqs. 6.6 and 6.10. In the presence of a magnetic field H, the off-diagonal components σ_{ij} are no longer zero, and the Hall mobility μ_H ($= R_H \sigma_{ii}$) is obtained from the value of $\sigma_{xy}^{(a)}/\sigma_{xx}H$, where we assume for simplicity that H is in the z direction, and the current is in the x direction. The quantity $\sigma_{xy}^{(a)}$ is the antisymmetric part of σ_{xy}, which in turn is expressed by Eq. 6.6 where now:

$$D_K(E) = \tfrac{1}{3} \int_0^\infty \langle v_x(0)v_y(\tau)\rangle_c \, d\tau. \qquad (6.19)$$

In contrast to the solution for the electrical conductivity, which depends on a matrix element between adjoining pairs of sites (Eq. 6.11), the present solution leads to an expression which depends on transfer integrals between three adjoining sites that form a loop enclosing the magnetic field. After carrying out averages with the help of some approximations, Friedman derives an expression:

$$\mu_H = 2\pi(ea^2/\hbar)(\eta\bar{z}/z)[a^3 JN(E)], \qquad (6.20)$$

where $\bar{z}$ is the average number of pairs of sites which form closed loops with respect to an arbitrary site, and η is a dimensionless parameter expressing the effect of the averaging process ($\eta \sim \frac{1}{3}$). The quantities J, a, and z are the same parameters which enter the theory for $\sigma(E)$ in Eqs. 6.12 and 6.13.

One important result of this theory is that it yields a negative sign for the Hall coefficient when the transport is by holes, and thus it provides an explanation of the sign discrepancy with the thermopower. This is closely connected with the assumption that three-site loops provide the dominant contribution to the evaluation of Eq. 6.19. Four-site loops would yield a positive Hall coefficient.

One notes that the Hall mobility in Eq. 6.20 is proportional to the first power of the density of states whereas the electrical conductivity varies as $[N(E)]^2$. For the range of σ where the metallic approximation is good, one can use Eqs. 6.11, 6.12, and 6.17 to write:

$$\mu_H = 0.134(\eta\bar{z}/z^{3/2})a^{*5/2}(\sigma/1000)^{1/2}(\text{cm}^2/\text{V sec}), \qquad (6.21)$$

where a^* is in angstroms and σ is in ohm^{-1} cm^{-1}. This indicates that for $\sigma \gtrsim 1000$ ohm^{-1} cm^{-1}, $\mu_H \gtrsim 0.5$ cm^2/V sec, in agreement with experimental results (see Fig. 2.12).

However, Eq. 6.21 also indicates that μ_H should vary as $\sigma^{1/2}$. This is in distinct disagreement with experimental results for some alloys in which σ has a strong dependence on T. We have noted in Chapter 2 that σ varies strongly with temperature in $Tl_x Te_{1-x}$ when $x < \frac{2}{3}$. Instead of changing also with T as $\sigma^{1/2}$, μ_H is constant (Donally and Cutler, 1972). The same thing is true for pure tellurium. In this case, as noted by Warren (1972a), Knight shift measurements show independently that $N(E_f)$ changes with T in accordance with the behavior inferred from $\sigma(T)$ (Eq. 6.14).

Thus Friedman's theory accounts for several important aspects of the Hall effect behavior in liquid semiconductors. But it still disagrees with experiment in some aspects, which suggests that some ingredients of the situation are not taken into account. This may have to do with the peculiar nature of the nearly filled band in p-type liquids.

6.3 THERMAL TRANSPORT

In Section 2.4 we have discussed the contributions to the thermal conductivity from atomic motion κ_a and radiation κ_r, and we need to consider here only the electronic contribution κ_e. The first question to be examined is whether the various transport mechanisms discussed in the preceding section cause special effects in κ_e, in particular the hopping and diffusive mechanisms. In Appendix A, an expression is derived for κ_e (Eq. A8) which is expressed in terms of a special average of $\sigma(E)$ which is analogous to the result for the thermopower (Eq. 6.3). This shows that the transport mechanism has no effect on κ_e which is not incorporated in $\sigma(E)$. This conclusion is subject to the starting assumptions, particularly that there is no nonelectronic heat of transport, as discussed in Section 6.1.1.

Equation A8 also shows that the Wiedemann–Franz law for metals is obeyed in general, since the metallic limit gives $\kappa_e/\sigma T = W_0 = (\pi^2/3)(k/e)^2$ regardless of the form of $\sigma(E)$. At the other limit, the Maxwell–Boltzmann approximation gives a value for W which depends on the form of $\sigma(E)$. In the frequently encountered case where $\sigma \propto E$, $W = 2(k/e)^2$. The situation in liquid semiconductors frequently lies between these two limits. We show in Fig. 6.2 a curve marked κ_b which describes the theoretical dependence of W on E_f/kT for the case where $\sigma \propto E$.

The failure to use accurate formulas for ambipolar transport has caused misleading or erroneous conclusions in a number of studies of thermal conductivity. Let us review first the phenomenological equations for two-band transport. The current density $\mathbf{J}$ and heat flux density $\mathbf{Q}$ caused by an electric field $\mathbf{E}$ and temperature gradient ∇T are given by:

$$\mathbf{J}_\alpha = \sigma_\alpha \mathbf{E} - S_\alpha \sigma_\alpha \nabla T, \tag{6.22}$$

$$\mathbf{Q}_\alpha = TS_\alpha \mathbf{J}_\alpha - \kappa_\alpha \nabla T, \tag{6.23}$$

where the subscript α refers to transport by electrons ($\alpha = n$) or by holes ($\alpha = p$). The condition $\mathbf{J} = \mathbf{J}_p + \mathbf{J}_n = 0$, gives:

$$\mathbf{E} = S \nabla T, \tag{6.24}$$

where:

$$S = (\sigma_n S_n + \sigma_p S_p)/(\sigma_n + \sigma_p) \tag{6.25}$$

is the ambipolar expression for S. Eliminating $\mathbf{E}$, $\mathbf{J}_n$, and $\mathbf{J}_p$, one gets for the total heat flux $\mathbf{Q} = \mathbf{Q}_p + \mathbf{Q}_n$:

$$\mathbf{Q} = -(\kappa_p + \kappa_n + \kappa_m) \nabla T, \tag{6.26}$$

where κ_m is the ambipolar thermal conductivity:

$$\kappa_m = \frac{\sigma_p \sigma_n}{\sigma_p + \sigma_n}(S_p - S_n)^2 T. \tag{6.27}$$

The MB approximation is made very frequently in discussions of ambipolar transport, which yields expressions for S_p and S_n similar to Eq. 6.5b. In this case, the factor $S_p - S_n$ in Eq. 6.27 has the form:

$$S_p - S_n = \frac{k}{e} \frac{A_p + A_n + E_G}{kT}, \tag{6.28}$$

where $E_G = E_c - E_v$ is the band gap, and A_p and A_n are constants whose value depends on the shape of $\sigma(E)$ at the two band edges. Since $\sigma_p \sigma_n/(\sigma_p + \sigma_n)^2$ is a fraction whose maximum value is $\frac{1}{4}$, $\kappa_m/\sigma T$ can be as large as $(k/e)^2(E_G/2kT)^2$. If $E_G \gg kT$, $\kappa_m/\sigma T$ will be a good deal larger than the value of W for monopolar transport.

This result has led to misleading inferences about liquid semiconductors. Since κ includes an atomic contribution κ_a (Eq. 2.3), a large value of $\kappa_m/\sigma T$ has significance only if $\kappa_m \gtrsim \kappa_a$. But σ itself becomes small in the MB limit, and in fact the range where $\kappa_m/\sigma T$ becomes large is one in which $\kappa_m \gtrsim \kappa_a$. In practical cases where σ is large enough so that $\kappa_e \gtrsim \kappa_a$, E_f must be close enough to the band edge so that the MB approximation is not valid. To see this, we recall the $\kappa_a \cong 4 \times 10^{-3}$ W/deg cm for liquids (Chapter 2.4), so that if $W \cong 2(k/e)^2$ and $T \sim 1000°K$, $\kappa_e \cong \kappa_a$ implies that $\sigma \sim 250$ ohm^{-1} cm^{-1}. Thus according to the discussion in Section 6.1.6, E_f must be near the mobility edge, and Fermi–Dirac integrals, discussed in Appendix B, must be used to determine κ_e in the range where ambipolar contributions are significant.

In order to judge the possible magnitude of the ambipolar contribution, we consider the simple situation where the conduction and valence bands are symmetrical about the band gap, and examine the behavior of $\kappa_e/\sigma T$ as a function of E_G/kT (Cutler, 1974b). The symmetrical situation maximizes ambipolar transport. It is necessary also to specify the shape of $\sigma(E)$, and we assume $\sigma = bE$, where b is constant, as has been found to be the case in Tl$_2$Te (Cutler, 1974a, 1976b). In Fig. 6.2 the dependence of $\kappa_e/\sigma T$ on $E_G/2kT$ (the curve labeled κ_b^*) is shown in comparison to $\kappa_a/\sigma T$ calculated with a typical value of b (2300 ohm^{-1} cm^{-1} eV^{-1}). The difference between this curve and the single band curve (marked κ_b) represents the ambipolar contribution. It is seen that the ambipolar effect is appreciable while the band gap is still negative, and it becomes very large for large E_G/kT as expected. However, $\kappa_a/\sigma T$ also becomes large and κ_a overtakes κ_b^* when $E_G \gtrsim 4kT$, whereas the MB approximation is inaccurate when $E_G < 8 \ kT$. When the bands

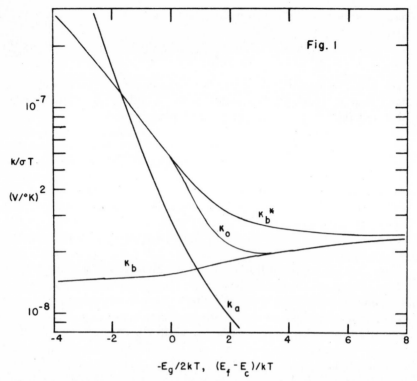

Fig. 6.2. Theoretical curves for $\kappa/\sigma T$ versus the Fermi energy $(=-E_G/2)$ normalized to kT for symmetrical overlapping parabolic bands. κ_b shows the single-band behavior, κ_o the two-band behavior, and $\kappa_b{}^*$ the two-band behavior with neglect of the extra contribution to $\sigma(E)$ due to the overlapping bands. κ_a indicates the behavior of $\kappa_a/\sigma T$ for a typical liquid semiconductor (Cutler, 1974b).

overlap, there is a contribution to $\sigma(E)$ in addition to the single band terms, which is discussed in Section 7.1 (Eq. 7.10). This term is neglected in $\kappa_b{}^*$ but is included in the curve marked κ_o.

6.4 MAGNETIC BEHAVIOR

6.4.1 Magnetic Susceptibility

The theory of magnetic susceptibility has been developed mainly in relation to other types of materials. Van Vleck's treatise (1932) treats the subject mainly in the context of nonmetallic materials, and a good presentation of the theory for metals has been made by Wilson (1954). Faber (1972), Busch and Yuan (1963), and Dupree and Seymour (Beer, 1972) have reviewed the subject in relation to liquid metals. As noted in Section 2.5, the magnetic susceptibility χ of liquid metals can be expressed as:

$$\chi = \chi_{\text{ion}} + \chi_P + \chi_L. \tag{6.29}$$

The Pauli paramagnetic term χ_P and the Landau diamagnetic term χ_L are derived from the free electron gas model to be:

$$\chi_P = -3\chi_L = \mu_B{}^2 N(E_f), \tag{6.30}$$

where μ_B is the Bohr magnetron (Busch, 1963). When the free electron model is refined to include the band structure and many-body effects, χ_P and χ_L behave somewhat differently. We refer the reader to Faber's discussion of these refinements (Faber, 1972).

In Section 2.5, a number of examples were discussed such as Te and In_2Te_3, where χ increases with T above the melting point, in parallel with a similar behavior for $\sigma(T)$. Equations 6.4a and 6.14 indicate that $\sigma \propto [N(E_f)]^2$, so that the parallel behavior of $\chi(T)$ and $\sigma(T)$ may be explained by an increase in $N(E_f)$ with T. Introducing this expression for σ into Eqs. 6.29 and 6.30 yields:

$$\chi = A + B\sigma^{1/2}, \tag{6.31}$$

where A and B are constants. This result is tested for tellurium and Tl–Te alloys in Fig. 6.3, where we plot χ versus $\sigma^{1/2}$. It is seen that the relationship is obeyed in the diffusive range of σ ($\gtrsim 2000$ ohm^{-1} cm^{-1}).

It is probably not accurate to attribute the diamagnetic term A in Eq. 6.31 to ions alone in the case of tellurium and other liquid semiconductors, since it is likely that they contain covalently bonded molecules. The diamagnetic susceptibility of both ions and molecules is expressed theoretically by the Langevin–Larmor formula:

$$\chi_{LL} = -(e^2/6mc^2) \sum_i \langle r_i{}^2 \rangle, \tag{6.32}$$

where $\langle r_i{}^2 \rangle$ is the mean square distance of a particular electron from the center of the molecule. The main contributions to the sum are from the more loosely bound valence electrons. Since the change in $N(E_f)$ which causes χ_P to increase with T may be caused by a change from covalent to metallic bonding, the question arises whether this may not cause χ_{LL} to also change with T, which would cause A in Eq. 6.31 to vary. We can only give a speculative answer to this question. If the diamagnetic term χ_L for a metal (Eq. 6.30) is expressed in terms of the radius r_e for the spherical volume per electron, using the free-electron expression for $N(E_f)$, it can be reduced to the same form as Eq. 6.32 with $\langle r_i{}^2 \rangle$ replaced by $1.22\, r_e{}^2$. This suggests that the diamagnetic term is not changed very much when covalent bonds

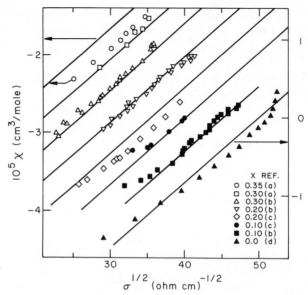

Fig. 6.3.   Plots of χ versus $\sigma^{1/2}$ for various compositions of Tl_xTe_{1-x}. The scales are given at the left and right for $x = 0.35$ and 0.00, respectively, and the data and reference line are displaced for intermediate compositions. The susceptibility measurements are by Gardner and Cutler (1976) and the conductivity data are from (a) Kazandzhan *et al.* (1972), (b) Lee (1971), (c) Enderby and Simmons (1969), and (d) Perron (1967).

change to metallic bonds, so perhaps the change in χ with T in liquid semiconductors can be attributed entirely to the change in Pauli term χ_P. Experimental evidence on this question is discussed in Section 7.5. A more direct test of this question can be made when χ_P can be determined from the Knight shift, as discussed in the next subsection. Warren (1972a) has discussed this problem in relation to the behavior of liquid Te. Faber's (1972) review of the behavior of χ in liquid metal alloys is pertinent to this problem.

In our discussion, we have neglected the van Vleck paramagnetic contribution to χ which can be described for a molecule as the effect of mixing of higher states under the perturbation of the magnetic field. The theoretical implications of this phenomenon for condensed systems such as crystals or liquids are complex, and it is not a very well developed subject. The reader is referred to a recent study of the problem by White (1974).

6.4.2   The Knight Shift

Nuclear magnetic resonance (NMR) frequencies and relaxation rates have been measured in a number of liquid semiconductors including Te (Cabane and Froidevaux, 1969; Warren, 1972a), Tl_xTe_{1-x} (Brown *et al.*, 1971; Warren,

1972b), In_2Te_3, Ga_2Te_3 and Sb_2Te_3 (Warren, 1971), Ga_xTe_{1-x} (Warren, 1972b), $Ga_2(Te_{1-x}Se_x)_3$ and Cu_xTe_{1-x} (Warren, 1973), and Se_xTe_{1-x} (Seymour and Brown, 1973). Important information has been derived from the Knight shift and from studies of the relaxation rate (discussed in the next subsection). In doing this, it is often necessary to consider fine points of theory and experiment which are not discussed here. The reader is referred to the original papers on these questions. A fundamental treatise on the subject has been written by Abragam (1961). Many aspects of the theory which are relevant to liquid metals and liquid semiconductors are reviewed by Warren and Clark (1969), Dupree and Seymour (Beer, 1972), and by Faber (1972). Bishop (1973) has reviewed the application of NMR to liquid and amorphous semiconductors.

Knight has shown that the hyperfine interaction between conduction band electrons in a metal and the nuclear magnetization leads to a relative change in the resonance frequency $K = \Delta\omega_N/\omega_N$:

$$K = (8\pi/3)\chi_P P_f \Omega, \qquad (6.33)$$

where Ω is the atomic volume, and P_f is $\langle|\Psi(0)|^2\rangle_f$, the probability density at the nucleus of the electron states at E_f. (Papers on NMR frequently refer to a density of states defined per atom and for a single electron spin, and normalize Ψ to the atomic volume rather than the total volume as we do here. We do not use these conventions in order to maintain consistency with other formulas in this book.)

Since $\chi_P \propto N(E_f)$, measurement of the Knight shift provides a more direct way to infer changes in $N(E_f)$ in liquid semiconductors than χ, which contains a diamagnetic term which is comparable to χ_P. However, there are several factors which may complicate the situation. Minor contributions to K from other processes besides the direct contact term given in Eq. 6.33 may have to be considered. The chemical shift may need to be determined, particularly when K is small. Finally, P_f must remain constant. The quantity P_f is determined primarily by the s component of the atomic part of Ψ, and it may change if the character of the wave functions at E_f changes with temperature or composition.

For liquids with $\sigma \gtrsim 200$ ohm^{-1} cm^{-1}, a constant P_f implies that:

$$K = \text{const } \sigma^{1/2}. \qquad (6.34)$$

This relationship has been found to be valid in tellurium (Mott and Davis, 1971; Warren, 1972a), Tl_xTe_{1-x} (Brown et al., 1971), and Se_xTe_{1-x} (Seymour and Brown, 1973). We show in Fig. 6.4 plots of ln K versus ln σ for a number of compositions of Tl_xTe_{1-x}, and it is seen that $K/\sigma^{1/2}$ is the same for a number of compositions. This indicates that P_f is independent of x in the range $0.4 \gtrsim x \gtrsim 0.6$, as well as independent of T. If the value of P_f is known,

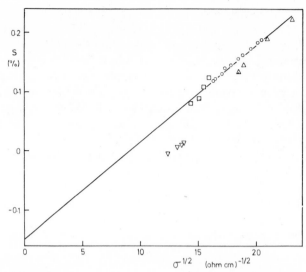

Fig. 6.4. ^{125}Te resonance shifts in Tl_xTe_{1-x} at various temperatures as a function of $\sigma^{1/2}$, for $x = 0.40$ (○), 0.55 (△), 0.60 (□), and 0.65 (▽). (Brown *et al.*, 1971).

it is possible to deduce the value of $N(E_f)$ from measurements of K. Warren (1971, 1973) has made estimates of P_f, and hence $N(E_f)$ for liquid Ga_2Te_3, In_2Te_3, and tellurium.

At the other extreme of possible behavior, P_f may change with composition or temperature while $N(E_f)$ remains approximately constant. Such a situation seems to occur in the metallic liquid alloy $In_{1-y}Bi_y$ (Styles, 1966). In Fig. 6.5, the Knight shifts of In and Bi are shown as a function of composition. There is a relatively small dependence on T, as expected for a metal. The functions $K(In)$ and $K(Bi)$ have a different dependence on the composition parameter y. This implies that P_f rather than χ_P is varying with composition, since the latter is common to both atoms. Styles has suggested that complexing occurs which reflects the tendency for compounds to occur at compositions In_2Bi and $InBi$. These compounds occur as solid phases at lower temperatures. Thus bismuth atoms have a constant environment ($\sim In_2Bi$) for $0 < y < \frac{1}{3}$ and $K(Bi)$ is constant, whereas indium atoms will have a constant environment ($\sim InBi$) in the range $\frac{1}{2} < y < 1$, causing $K(In)$ to be constant. If clustering causes a modification of the wave function at the nuclear sites (particularly the s component), the value of P_f can be identified with the partial covalent binding which causes the cluster. It should be noted that the lifetime of the clusters is expected to be small compared to the NMR period ω_N^{-1} so that the value of P_f which goes into Eq. 6.33 is an average of the values observed at the nucleus for its various configu-

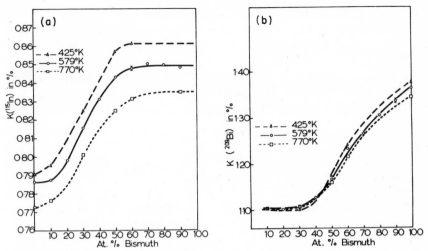

Fig. 6.5. Knight shift isotherms in In–Bi alloys (a) ^{115}In and (b) ^{209}Bi; 425°K (———),
579°K (———), 770°K (----) (Styles, 1967).

rations. Otherwise one would observe different resonance frequencies for
different clusters.

In general it is possible for both P_f and $N(E_f)$ to vary simultaneously.
An interesting example is provided by $Ga_{1-y}Te_y$, which is discussed in
Section 8.5.

6.4.3   Nuclear Relaxation Rate

The nuclear relaxation rate R is the sum of the rates R_M due to magnetic
interactions and R_Q due to quadrupole interactions. Interesting information
has been derived about liquid semiconductors from both of these parameters.
In many nuclei R_Q is zero because of the absence of a quadrupole moment.
(This happens when the nuclear spin $I < 1$.) If the two rates are comparable,
it is sometimes possible to separate them by comparing the total rates R for
two isotopes of the same nucleus (Warren and Clark, 1969; Warren, 1971).

In metallic situations, the direct interaction between the electron and the
nuclear spin provides an important mechanism for magnetic relaxation. The
Korringa theory applies for R_M in the case of an electron gas, according to
which:

$$(R_M)_K = (4\pi\gamma_n^2 kT/\gamma_e^2\hbar)K^2, \tag{6.35}$$

where γ_n and γ_e are the nuclear and electron gyromagnetic ratios, respectively.
Warren (1971) has derived a theory for the behavior of R_M when electron
scattering is strong as in the diffusive domain, which we now outline. It

begins with the observation that R_M is proportional to the decay constant τ_e for time-dependent correlations of the matrix element between the nuclear and electron spins. In normal metals, τ_e is identified with the ratio of the Fermi velocity to the interatomic distance. But when scattering is strong, τ_e is the same as v_e^{-1} where v_e is the frequency factor in $D_K(E)$ defined in Eq. 6.9. When an arbitrary value of τ_e replaces the one derived for gas-like motion, R_M is increased over the Korringa value $(R_M)_K$ in Eq. 6.35 by a factor η_R:

$$\eta_R = R_M/(R_M)_K = \tau_e/[\hbar\Omega N(E_f)/2], \qquad (6.36)$$

where the term in the square bracket represents approximately the value of τ_e in the free-electron approximation.

Equation 6.15 describes the enhancement of $\tau_e (= v_e^{-1})$ which occurs in diffusive transport theory when $N(E)$ is small. Equation 6.36 provides a direct measure of the decrease in v_e, and substituting it into the expression $\sigma(E) = (a^2/6\tau_e)N(E)$ leads to Warren's relationship:

$$\eta_R\sigma = \sigma_0 \cong \mathrm{const}(e^2/\hbar a), \qquad (6.37)$$

where the constant is of the order of magnitude of unity.

According to this, η_R should vary as σ^{-1} in the diffusive range. When σ increases beyond σ_0, the Korringa relation becomes valid, and η_R should become constant and equal to 1. We show in Fig. 6.6 Warren's plot of log η_R versus log σ for a number of conducting liquids. In many liquids with larger σ, $\eta_R \sim 1$. In liquid In_2Te_3 and Ga_2Te_3, on the other hand, σ varies with T in a range below 2500 ohm^{-1} cm^{-1} and $\eta_R \propto \sigma^{-1}$ with $\sigma_0 \cong 1800$

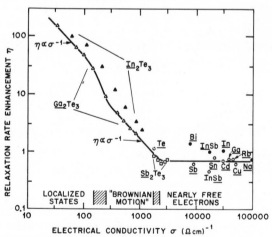

Fig. 6.6. Magnetic relaxation enhancement η versus electrical conductivity σ for liquid metals and semiconductor. The closed points are affected by quadrupole contributions to the relaxation rate (Warren, 1971).

ohm^{-1} cm^{-1}. This result provides direct support for the validity of the diffusive model. It also confirms the estimated upper limit for the diffusive range $\sigma(E) \gtrsim 2500$ ohm^{-1} cm^{-1}. In liquid Te, Warren (1972b) has found a deviation from $\eta_R \propto \sigma^{-1}$ in the low temperature range $1000 < \sigma < 1500$ ohm^{-1} cm^{-1}. This suggests that other factors may be playing a role in K or R_M which are not taken into account in Eq. 6.36.

The quadrupole relaxation rate R_Q depends on an interaction between the nuclear quadrupole moment and an electric field gradient at the nucleus. The latter is generally very small in a liquid metal, but it may be rather large in a covalent molecule or a crystal for which the electrostatic field at the nuclear site is not symmetrical. In a liquid, asymmetric fields which occur normally persist over vibrational time intervals τ_v ($\sim 10^{-12}$ sec) which are short compared to ω_N^{-1} ($\sim 10^{-7}$ sec), so that R_Q is small because of time averaging. Therefore relatively large values of R_Q indicate molecular bonding configurations which persist over times which are long compared to τ_v. Warren (1971) has deduced a value of the configuration correlation time τ_c ($\sim 10^{-11}$ sec) for Ga_2Te_3 near the melting point, which is an order of magnitude greater than typical values for liquid metals. There are difficulties in evaluating the magnitude of τ_c because of uncertainties in the antiscreening factors caused by the polarization of the electronic shell surrounding the nucleus (Kerlin and Clark, 1975). Therefore, a change in R_Q with temperature is perhaps more strongly indicative of bonding effects. Styles (1966) has observed a strong increase in R with decreasing T for bismuth in liquid In_xBi_{1-x} at $x = 0.5$. Although he did not resolve R into the magnetic and quadrupole contributions, a likely mechanism seems to be an enhancement of R_Q at low T. This is caused by an increasing correlation time for the molecular clusters.

6.5   OPTICAL PROPERTIES

Study of the absorption edge for optical transitions between the valence and conduction bands has been the classic method for determining the band gap in crystalline semiconductors. Few liquid semiconductors have been studied in this way. These include As_2Se_3 and related compounds (Edmond, 1966) and Se_xTe_{1-x} alloys (Perron, 1972). We show in Fig. 6.7 the absorption curves for As_2Se_3 at a number of temperatures both above and below the glass transition temperature. In order to determine the exact size of the band gap from the optical absorption coefficient α versus frequency ω, it is necessary to have an adequate theory for shape of the absorption edge. This is lacking for disordered materials. Reviews of this subject have been written by Mott and Davis (1971), Tauc and Menth (1972), and by Tauc (1974). The role played by the localized states in the pseudogap between the mobility edges presents a special problem for interpreting the spectrum of disordered materials.

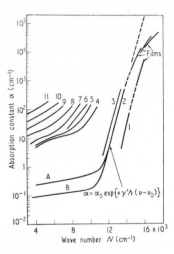

Fig. 6.7. Optical absorption curves for As_2Se_3 in the glassy and liquid ($T > 200°C$) states; (1) $-196°C$, (2) 24°C, (3) 80°C, (4) 288°C, (5) 349°C, (6) 386°C, (7) 438°C, (8) 478°C, (9) 524°C, (10) 554°C, (11) 597°C; $v = cN$, $\gamma' = $ constant (Edmond, 1966).

The $\alpha(\omega)$ curves for As_2Se_3 are exponential, in keeping with the Urbach rule, but they deviate from that rule in that $d \ln \alpha/d\omega$ is not proportional to T^{-1}. Similar behavior has been observed in liquid Te–Se alloys. The theoretical basis of the Urbach rule is uncertain, although an explanation by Dow and Redfield (1972) in terms of exciton line broadening by random electric fields seems promising. Mott and Davis (1971a) suggest that the As_2Se_3 curves may reflect free carrier absorption, which is discussed below. Edmond's results for $\sigma(\omega)$ of liquid $Tl_2Te \cdot As_2Te_3$ are qualitatively similar to As_2Se_3. In this case measurements by Mitchell, Taylor, and Bishop (1971) on the vitreous phase has produced evidence linking the mechanism for $\sigma(\omega)$ between DC and optical frequencies ($5-50\mu$).

In the absence of a reliable theory for the absorption edge, a relatively crude method must be used for deducing the dependence of the band gap on temperature, such as determining the value of ω at which α has some arbitrary value ($\alpha = 1000$ cm^{-1} is frequently chosen for this purpose). Using this procedure, Edmond (1966) has derived the temperature dependence of the band gap E_G. It falls from ~ 2 eV at low temperatures (glassy state) to values approaching zero at $\sim 600°C$. The quantity $-dE_G/dT$ is rather large ($\sim 1.6 \times 10^{-3}$ eV/deg), and this is in accord with information derived from electrical transport measurements. It is an order of magnitude larger than the typical values for crystalline solids, but similar large values of $-dE_G/dT$ have also been found for a number of amorphous solids and other liquid semiconductors.

Most of the well-studied liquid semiconductors are highly conducting ($\sigma \gtrsim 100$ ohm^{-1} cm^{-1}). For these materials free-carrier absorption must be important, and α is too large to be measured except by reflection or emission methods (Tauc and Abraham, 1968). Measurements have been reported for molten Te (Hodgson 1963) and for molten CdTe (Tauc and Abraham, 1968) which indicate that the optical frequency electrical conductivity $\sigma(\omega)$ increases with ω by a factor ~ 2 with increasing ω.

The theory for free-carrier absorption in the diffusive range of transport has been derived by Mott and Davis (1971) and by Hindley (1970). The result is

$$\sigma(\omega) = \int_{-\infty}^{\infty} \left[\left(\frac{2\pi\hbar^3 a\lambda^2 e^2}{3m^2} \right) N(E)N(E + \hbar\omega) \right] \left[\frac{f(E) - f(E + \hbar\omega)}{\hbar\omega} \right] dE. \quad (6.38)$$

It is illuminating to compare the form of Eq. 6.38 with Eq. 6.1 for the DC electrical conductivity σ. Reference to Eq. 6.14 shows that the quantity in the first bracket is the geometric mean of $\sigma(E)$ and $\sigma(E + \hbar\omega)$ which we designate $\sigma(E, \hbar\omega)$. The quantity in the second bracket is analogous to $-\partial f/\partial E$ shown in Fig. 6.1a in that it has a unit area, but it is rectangular in shape, extending from $E_f - \hbar\omega$ to E_f. Thus, the integral in Eq. 6.38 corresponds to the average of $\sigma(E, \hbar\omega)$ between E_f and $E_f - \hbar\omega$. If $\sigma(E)$ varies moderately with E in the vicinity of E_f (say according to a power law), $\sigma(\omega)$ can be expected not to differ greatly from σ over a moderate range of ω. If the density of states drops to zero at an energy interval E_1 above or below E_f, one would expect $\sigma(\omega)$ to decrease as ω^{-1} at $\hbar\omega \gtrsim E_1$. On the other hand, if the density of states increases abruptly at an interval $\pm E_1$, as the result of another band or because E_f is in the region of a minimum in the density of states, then $\sigma(\omega)$ should start to increase for $\hbar\omega \gtrsim E_1$.

These remarks ignore the possible effects of localized states or states with gaslike transport properties which might occur in the range $E_f + \hbar\omega$. Hindley has shown that if the initial or final states for an optical transition are within the mobility gap, their contribution to $\sigma(\omega)$ would still be in accord with Eq. 6.14 as long as the states at the other end of the transition are nonlocalized. As for states in which gaslike motion occurs, we expect that the qualitative conclusions implied by Eq. 6.38 would still be applicable. The reader is referred to discussions of the optical behavior of liquid metals by Faber (1972) or Hodgson (Beer, 1972) for more information on this question.

In the case of molten Te (Hodgson, 1963), $\sigma(\omega)$ rises by a factor ~ 2 in the range of $\hbar\omega \sim 0.3$ to 1.2 eV. This suggests that $N(E)$ and $\sigma(E)$ rise abruptly at an energy interval ~ 0.3 eV above or below E_f. In the case of CdTe (Tauc and Abraham, 1968), a rise in reflectivity occurs for $\hbar\omega \gtrsim 1$ eV, which puts an extra band at that distance from E_f.

The reader will note that inferences have been made about interband absorption from a formula which was derived for intraband (free carrier) absorption. This is justified by the random phase model used in its derivation. Since $\sigma(E)$ is derived from interactions between tight binding states at adjoining sites, the only effect of a different band index is for one of them to modify the value of λ in Eq. 6.14.

The quantity α (in cm^{-1}) is related to $\sigma(\omega)$ (in $ohm^{-1} cm^{-1}$) by:

$$\alpha(\omega) = (120\pi/n)\sigma(\omega), \tag{6.39}$$

where n is the refractive index. Taking $n = 3$ as a typical value, $\alpha \sim 100\sigma$. Unless there is a wide band gap, $\sigma(\omega)$ is generally larger than $\sigma(0)$, so that an experimental limitation $\alpha \gtrsim 1000 \, cm^{-1}$ for transmission measurements implies that absorption measurements are not feasible for liquids whose DC conductivity exceeds $\sim 10 \, ohm^{-1} \, cm^{-1}$. That explains the absence of optical absorption measurements for many liquid semiconductors.

In addition to information about the band gap, optical measurements provide a possible means for obtaining information about chemical bonding. Infrared absorption lines have been observed in glassy As_2Se_3 which persist into the liquid range of temperatures. The lines disappear abruptly at 120°C above the vitrification temperature, and this has been taken as evidence that the bond network structure breaks up within a small range of temperature (Taylor *et al.*, 1971). Absorption lines can of course be observed only in high-resistivity liquids. Raman scattering can also be studied in high-resistivity liquids, and information about molecular vibrations has been obtained for As_2S_3 (Finkman *et al.*, 1973) and As_2Se_3 (Finkman *et al.*, 1974).

CHAPTER 7

Tl–Te ALLOYS

In earlier chapters we have reviewed information about liquid semiconductors on both empirical and theoretical levels. In the remaining chapters we examine a number of specific systems for how well experimental information can be interpreted coherently in terms of the available theoretical concepts and relations. The Tl–Te system discussed in the present chapter provides a natural starting point because there is a large body of good experimental data, and significant progress has been made in interpreting the electronic behavior of this system. Since many of the facets of behavior of Tl–Te are exhibited in other liquids, the results obtained for Tl–Te may be expected to provide a useful guide for understanding the other less thoroughly studied systems. These other systems will be considered in the final chapter.

Some aspects of the behavior of Tl–Te have already been discussed in earlier chapters; we first review a general model for Tl–Te which is suggested by this information. The occurrence of a compound Tl_2Te is indicated by the thermochemical data (Figs. 3.5 and 3.6) and by the shape of the isotherms for the resistivity and thermopower (Fig. 2.3). On the Tl-rich side of this composition ($x > \frac{2}{3}$ in Tl_xTe_{1-x}), σ increases rapidly with x, and S is negative and decreases rapidly in magnitude. At the same time, both parameters have small temperature coefficients. This behavior, which we have previously categorized as M-type, suggests that the Tl atoms in excess of Tl_2Te are ionic, so that their valence electrons are added to the conduction band. The character of the conduction band as it is affected by the change in composition from Tl_2Te to Tl is discussed in Section 7.2. In the vicinity of the composition Tl_2Te, the conduction band is due to the Tl_2Te molecules.

At $x = \frac{2}{3}$, the alloy is expected to consist of Tl_2Te molecules plus relatively small amounts of dissociation products or other molecular species (see Section 3.4). The minimum conductivity at this point and the small thermopower relative to neighboring compositions indicates that the Fermi energy is near the middle of a pseudogap, and electronic transport occurs in two bands. This conclusion is supported by the behavior of the Hall mobility, discussed in Section 2.3.

When Te is added to Tl_2Te, σ increases relatively slowly, and S becomes positive and decreases slowly with x. Both quantities now exhibit a relatively large temperature coefficient. (See Figs. 2.2 and 2.8). This behavior has been characterized as S-type. Viewing this as an effect of doping of Tl_2Te with Te, most of the excess Te atoms must be electrically neutral. A small, strongly temperature-dependent fraction of the excess Te atoms is in an ionic form. Since S is positive, transport is due to holes, and the ions must be negative. We have described, in Section 5.4, a model for the electronic structure which could cause this behavior. The excess Te atoms combine with Tl_2Te to form chain molecules of the form $Tl–Te_n–Tl$, and some of the Te–Te bonds dissociate thermally to form $Tl–Te_n$ fragments plus holes in the valence band. Increasing the Te concentration increases the number of Te–Te bonds and the hole concentration at a given T.

The model outlined above is the basis for the discussions which follow. Other models have been proposed for Tl–Te, usually with only qualitative explanations of the experimental behavior. Hodgkinson (1971, 1973) and Cohen and coworkers (1972, 1973) have interpreted $Tl_x–Te_{1-x}$ data for $x < \frac{2}{3}$ in terms of heterogeneous transport. As pointed out in Section 3.4, we think that thermodynamic evidence precludes large enough clusters to justify heterogeneous transport theory. Ratti and Bhatia (1975) have analyzed the behavior of σ and R_H in terms of a theory for the electronic behavior proposed by Enderby and coworkers (1969, 1970), deriving the electron density from thermochemical data. However they note that the behavior of the thermopower is inconsistent with this theory for $x < \frac{2}{3}$. More general models for the electronic structure of Tl–Te and other alloys have been introduced by Faber (1972) and Roth (1975) as well as the one already mentioned due to Enderby and coworkers. These models are discussed more appropriately in Section 8.1.

7.1 M-TYPE RANGE: $\frac{2}{3} < x < 0.7$

7.1.1 Single-Band Model

At the composition Tl_2Te we expect the band structure to be given by Fig. 5.4b with the difference that the bonding and antibonding bands caused by Te–Te bond orbitals would be missing. Another possibility, to be discussed later, is that the conduction band is generated from nonbonding 6p orbitals of the Tl atoms in the Tl_2Te. Addition of excess Tl then introduces ions plus electrons which go into the conduction band, causing E_f to increase. Since $100 \gtrsim \sigma \gtrsim 2000$ ohm^{-1} cm^{-1} for $x < 0.72$, transport is in the diffusive range where $\sigma(E) \propto [N(E)]^2$. For this limited range of composition, it is reasonable to expect a rigid band model to be valid, i.e., $N(E)$ and hence $\sigma(E)$ is invariant when x or T is changed. Work by Cutler and Field (1968) has in fact shown that this is so, and that $\sigma \propto E$. We derive this result below in terms of current theory (Cutler, 1974a).

In Appendix B, we discuss the expression of σ, S, and the electron density n in terms of Fermi–Dirac integrals F_n. For the present case:

$$N(E) = C_n(E - E_{c0})^{1/2}, \tag{7.1}$$

where C_n is a constant. According to Eq. 6.14,

$$\sigma = AC_n^2(E - E_{c0}). \tag{7.2}$$

This yields:

$$n = C_n(kT)^{3/2}F_{1/2}(\xi), \tag{7.3}$$

$$\sigma = AC_n^2 kT F_0(\xi), \tag{7.4}$$

and:

$$S = -(k/e)[2F_1(\xi)/F_0(\xi) - \xi], \tag{7.5}$$

where:

$$\xi = (E_f - E_{c0})/kT. \tag{7.6}$$

If the Tl ions have a charge ze, there are z electrons in the conduction band for every Tl atom in excess of the composition Tl_2Te. This leads to an electron density:

$$n_0 = 3zN_a(x - \tfrac{2}{3}), \tag{7.7}$$

where N_a is the concentration of atoms. Density measurements yield a value $N_a = 2.7 \times 10^{22}$ cm^{-3} for Tl_2Te (Nakamura and Shimoji, 1973).

For a rigid band, there is an implicit relation between σ and S at a constant temperature through their mutual dependence on ξ in Eqs. 7.4 and 7.5, conveniently represented by plotting $\ln F_0$ versus S. This is compared with an experimental plot of $\ln \sigma$ versus S in Fig. 7.1, where the points represent various compositions of Tl_xTe_{1-x} at $T = 800°K$. The heavy curve is the theoretical relation derived from Eqs. 7.4 and 7.5, with the factor $AC_n^2 kT = 197$ ohm^{-1} cm^{-1}. The lighter curves are the best fits for theoretical curves derived for $r = 0.5$ and 1.5 in $\sigma \propto E^r$. (The latter are described by equations similar to Eqs. 7.4 and 7.5, as shown in Appendix B.) It is seen that there is a very good fit to the curve for $r = 1$ and we infer from this that the band edge is nearly parabolic.

It is also possible to calculate theoretical curves for $\ln \sigma$ versus S which allow for the existence of a mobility edge. This is done with the aid of modified

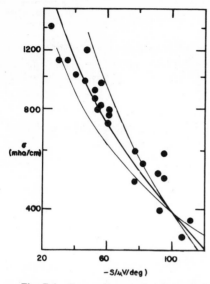

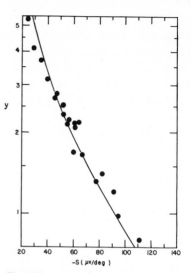

Fig. 7.1. Comparison of theoretical curves with experimental data for ln σ versus S in $Tl_2Te + Tl$ at $T = 800°K$. The heavy line is for $r = 1$ in $\sigma = $ const E^r, and the light lines are for $r = 0.5$ and 1.5 (Cutler and Field, 1968).

Fig. 7.2. Comparison of the theoretical curve with experimental data for composition of Tl_xTe_{1-x} versus the thermopower at $T = 800°K$. The parameter $y = 100(x - \frac{2}{3})$ (Cutler and Field, 1968).

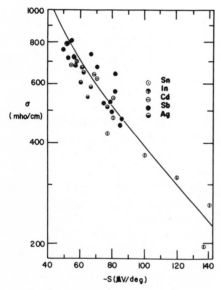

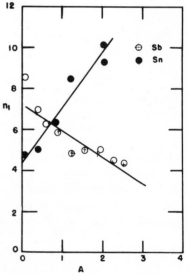

Fig. 7.3. Comparison of theoretical curve with experimental data for ln σ versus S at $T = 800°K$ in Tl–Te alloys doped with various metals. The theoretical curve is the same ($r = 1$) as in Fig. 7.1 (Cutler and Field, 1968).

Fig. 7.4. Dependence of the electron concentration, with an arbitrary scale, on the concentration A of Sn or Sb added to Tl–Te alloys (Cutler and Field, 1968).

Fermi–Dirac integrals containing an additional parameter $x_c = (E_{c1} - E_{c0})/ kT$, where E_{c1} is an energy below which $\sigma(E)$ is assumed to drop abruptly to zero. The equations for the mobility edge model are discussed in Appendix B3. The theoretical curves for $\ln \sigma$ versus S for various values of x_c are shown in Fig. B3, and it is seen that the shape is changed significantly when $x_c \gtrsim 2$. The shape of the experimental curve in Fig. 7.1 indicates that the distance of the mobility edge from the band edge is not larger than kT (~ 0.07 eV).

Equations 7.3, 7.5 and 7.7 describe the dependence of S on composition. Since $F_{1/2}(\xi)$ is proportional to $(x - \frac{2}{3})$, the theoretical relationship to composition was tested by plotting $\ln(x - \frac{2}{3})$ versus S at $T = 800°$K, and comparing this with a theoretical plot of $\ln F_{1/2}$ versus S. This is shown in Fig. 7.2, and again there is very good agreement with $C_n/z = 1.49 \times 10^{22}$ eV$^{-3/2}$ cm^{-3}.

Cutler and Field (1968) also investigated the effects of adding small concentrations of a third element such as Ag or Sn to Tl$_x$Te$_{1-x}$ alloys with $x \gtrsim \frac{2}{3}$. With minor exceptions, $\sigma(T)$ and $S(T)$ had the same shape as the Tl–Te curves. Furthermore, a plot of $\ln \sigma$ versus S at $T = 800°$K yields the same curve as the Tl–Te data. This is shown in Fig. 7.3, and the theoretical curve fits best with a somewhat smaller value of $AC_n{}^2 kT = 169$ ohm^{-1} cm^{-1}. The average yields the value $AC_n{}^2 = 2640$ ohm^{-1} cm^{-1} eV^{-1}.

These results show that addition of a small concentration c of a third element has the effect of changing n, and one expects a change in the electron density $\Delta n = cz_M$, where z_M is the valence of the metal M. Using the theoretical curve in Fig. 7.2, it is possible to calculate n/C_n as a function of c from the experimental value of S. In Fig. 7.4, this is plotted for the impurities Sn and Sb. Similar curves were obtained for all of the impurities investigated. The slopes of these curves provided values of z_M/C_n which were compared to the value for Tl from Fig. 7.2. It was found that these numbers are commensurate with a set of integers for z_M, with $z = 1$ for Tl, as shown in Table 7.1. It is seen that they correspond to common chemical valences for the metals: Ag(1), Cd(2), In(1), and Sn(2). Sb, with a valence -1, apparently forms a singly-charged negative ion. An error in the original work led to the false conclusion that $z = 3$ for thallium, and this was corrected in a later paper which discusses the chemical aspects in some detail (Cutler, 1973). In particular, it is noted that the results do not generally indicate whether the bonding of the third element is ionic or covalent.

7.1.2 Overlapping Band Model

We have considered so far only data at a single temperature 800°K. Assuming that n is the same as n_0 in Eq. 7.7, Eqs. 7.3, 7.4 and 7.5 also predict the effects of T on S and σ. The theoretical and experimental curves are

TABLE 7.1

Valences of Impurities for
Thallium Chalcogenides Tl$_2$X with
Excess Metal[a]

	Valence					
X	Tl	Ag	Cd	In	Sn	Sb
Te[b]	1	1	2	1	2	−1
Se[c]	1	1	2	3		
S[d]	1	1	2	3		

[a] Cutler (1973).
[b] Cutler and Field (1968).
[c] Nakamura and Shimoji (1969).
[d] Nakamura *et al.* (1974).

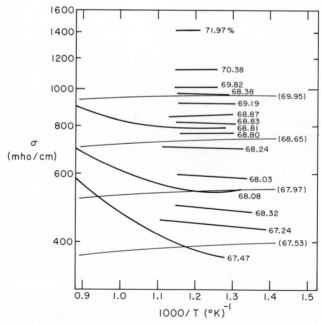

Fig. 7.5. Experimental curves (heavy lines) and single-band-model theoretical curves
(light lines) for $\sigma(T)$ of Tl$_x$Te$_{1-x}$ for $x > \frac{2}{3}$. The compositions are given in at% Tl (Cutler and
Field, 1968).

compared in Figs. 7.5 and 7.6. It is seen that there is a modest discrepancy which increases with T and as $x \to \frac{2}{3}$. This is what is expected qualitatively as the result of excitation of electron-hole pairs across a band gap. If the hole contribution to transport can be neglected because of trapping in localized states between the band edge E_{v0} and the mobility edge E_{v1} of the valence band, σ and S are still related by Eqs. 7.4 and 7.5, but one would now plot $800\ \sigma/T$ instead of σ. This is indeed found to be the case except for data at $T \gtrsim 1000^\circ$K. Consequently it was possible to determine the hole concentration $p = n - n_0$ as a function of T by means of Eqs. 7.3 and 7.4, and this was analyzed in terms of a simple two-band model for the pseudogap. Assuming, somewhat arbitrarily, that the valence band edge is parabolic, so that the density of states in the valence band $N_v = C_p(E_{v0} - E)^{1/2}$, the hole concentration is:

$$p = C_p(kT)^{3/2} \exp[-\xi - E_{G0}/kT] \exp[E_{G1}/k]. \tag{7.8}$$

The Maxwell–Boltzmann approximation used in this expression is good as long as $E_f - E_{v0} \gg kT$. We have made the common assumption that the band gap E_G varies linearly with T, so that:

$$E_G = E_{c0} - E_{v0} = E_{G0} - E_{G1}T. \tag{7.9}$$

Equation 7.8 suggests that one should plot $\ln(pe^\xi/T^{3/2})(\equiv f_p)$ versus T^{-1} to get E_{G0}; such plots are shown in Fig. 7.7. The lower curves with the large slopes correspond to the assumptions of our model and lead to an average value $E_{G0} = 0.58$ eV. The other curves correspond to situations where E_G

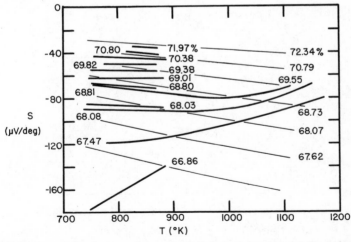

Fig. 7.6. Experimental curves (heavy lines) and single-band-model theoretical curves (light lines) for $S(T)$ of Tl$_x$Te$_{1-x}$ for $x > \frac{2}{3}$. The compositions are given in at% Tl (Cutler and Field, 1968).

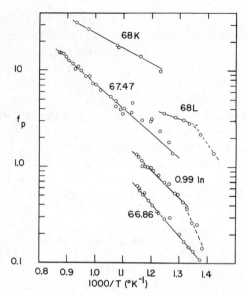

Fig. 7.7.   Plots of $\ln f_{\mathrm{p}}$ versus T^{-1} for several n-type Tl–Te alloys, where $f_{\mathrm{p}} = p\exp(\xi)/T^{3/2}$ (Cutler, 1974a).

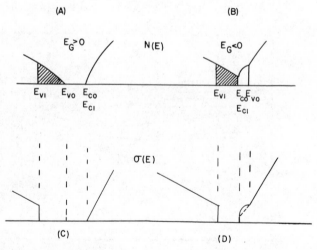

Fig. 7.8.   Two-band-model density of states for a (a) positive and (b) negative band gap. The localized states are indicated by hatching. The corresponding curves for $\sigma(E)$ are shown in (c) and (d). The part above the dashed line in (d) is due to the cross-term $2N_{\mathrm{v}}N_{\mathrm{c}}$ in $[N(E)]^2$ (Cutler, 1974a).

itself is negative because of the large value of E_{G1}. Several curves such as the ones marked 0.99 In and 68 L show a break at $T_0 = 770°$K. This has been interpreted to be the point where the bands overlap, and it leads to a value $E_{G1} = E_{G0}/T_0 = 7.5 \times 10^{-4}$ eV/deg.

When the valence band overlaps the conduction band, there will be an increase in $\sigma(E)$ at $E > E_{c0}$ as the result of two mechanisms as illustrated in Fig. 7.8: (1) In view of Mott's observation that there must be a sharp boundary between localized and extended states (Section 5.4), any localized states in the valence band must become nonlocalized when their energies exceed E_{c0} ($= E_{c1}$). Therefore these valence band states will become conducting, contributing a term $A[N_v(E)]^2$ to $\sigma(E)$. (2) In addition to the terms AN_c^2 and AN_v^2 associated with the individual bands, Eq. 6.14 yields a mixed term $2AN_vN_c$ when $N(E) = N_c + N_v$, which makes a further contribution to $\sigma(E)$ in the overlap region.

These considerations lead to an "overlapping band" model for σ and S which neglects contributions to $\sigma(E)$ from the part of the valance band for which $E < E_{c0}$:

$$\sigma(E) = AC_n^2(E - E_{c0}) \quad \text{for} \quad E > E_{v0},$$
$$= A[C_n^2(E - E_{c0}) + 2C_nC_p(E - E_{c0})^{1/2}(E_{v0} - E)^{1/2} + C_p^2(E_{v0} - E)]$$
$$\text{for} \quad E_{c0} < E < E_{v0},$$
$$= 0 \quad \text{for} \quad E < E_{c0}. \tag{7.10}$$

In order to apply this model, it is necessary to specify one further parameter, C_p. This was obtained by matching the theoretical curve for $\sigma(T)$ to a single experimental curve (at $x = 0.680$), which yielded $C_p = 0.42\ C_n$. With these parameters, a set of theoretical curves for $\sigma(x, T)$ and $S(x, T)$ are obtained which are compared with the experimental curves in Figs. 7.9 and 7.10. (A further adjustment was made which corresponds to reducing the values of C_p and C_n by 10%, in order to allow for the small effect of the overlapping bands in the original evaluation of AC_n^2 at $T = 800°$K.) We refer the reader to the original paper (Cutler, 1974a) for details on the method of calculation. There is a marked improvement from the results in Figs. 7.5 and 7.6. The fact that the $S(T)$ curves agree well with experiment, whereas the values of C_p, E_{G0} and E_{G1} were inferred solely from analysis of the $\sigma(T)$ curves, argues very strongly for the validity of the model.

The overlapping band model completely ignores ambipolar effects, that is, transport contributions from the valence band at energies below E_{c0}. Therefore, it is not suprising that there are increasing discrepancies with experimental curves at compositions approaching Tl_2Te. Curve A in Fig. 7.11 is the theoretical curve for $\sigma(T)$ at $x = 0.6667$, and it differs by $\sim 20\%$ from the experimental curve. The latter, due to Kazandzhan and coworkers

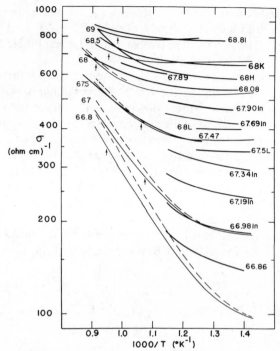

Fig. 7.9.   Comparison of theoretical curves for the overlapping band model (indicated by arrows) with experimental curves for $\sigma(T)$. The labels on the experimental curves indicate the composition in at% Tl, except for In-doped alloys marked In, where the equivalent composition is given in terms of pure Tl–Te. The dashed curves are theoretical curves containing a correction for ambipolar transport (Cutler, 1974a).

(Kazandzhan *et al.*, 1972) is unusually precise, and contains an interesting inflection which is small enough to normally be ignored. But comparison with curve A (where the inflection is emphasized by dashed extensions of the linear curves at lower and higher temperatures) indicates that it is due to the disappearance of the band gap. The corresponding theoretical curve A for $S(T)$ at the composition Tl_2Te is compared with experimental data in Fig. 7.12. There is some uncertainty in the experimental curve at lower temperatures due to the great sensitivity to small deviations in stoichiometry, as indicated by the two experimental curves with slightly different compositions. But the theoretical curve A lies far from either experimental curve. The greater sensitivity of the thermopower to ambipolar effects is to be expected, and it shows up also at $x = 0.6686$ in Fig. 7.10.

The ambipolar corrections to $\sigma(T)$ and $S(T)$ are sensitive to details about the shape of the valence band edge and the position of the mobility edge E_{vl}.

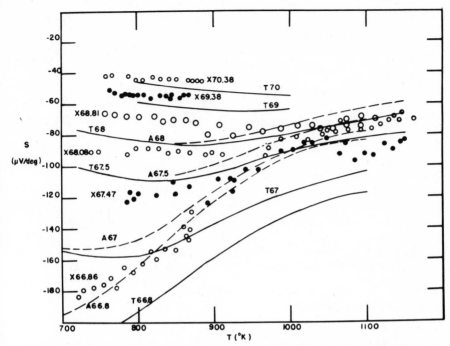

Fig. 7.10. Comparison of theory and experiment for $S(T)$ with the overlapping band model. Compositions are given in at% Tl, and are preceded by X for experimental curves (data points), T for theoretical curves (solid lines) and A for theoretical curves with an ambipolar correction (dashed lines) (Cutler, 1974a).

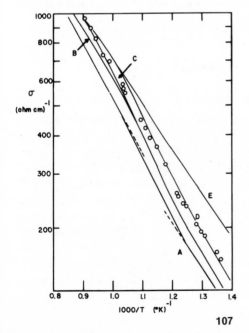

Fig. 7.11. Comparison of the overlapping-band-model theoretical curve (A) for $\sigma(T)$ at $x = \frac{2}{3}$ with experimental points (from Kazandzhan *et al.*, 1972). The curves marked B, C, D, and E contain ambipolar corrections according to various models for the valence band edge (Culter, 1974a).

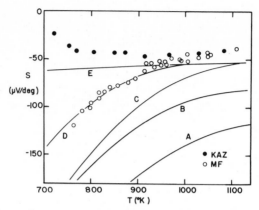

Fig. 7.12.   Comparison of the overlapping-band-model theoretical curve (A) for $S(T)$ at $x = \frac{2}{3}$ with experimental points. The closed circles are data points from Kazandzhan (1972 *et al.*, and the open circles are for a slightly different composition from Field (1967). The curves marked B, C, D, and E contain ambipolar corrections according to various models for the valence band edge (Cutler, 1974a).

Curves $\sigma(T)$ and $S(T)$ are shown in Figs. 7.10, 7.11, and 7.12 which incorporate ambipolar corrections following several reasonable models for the valence band. Although some of them remove most of the discrepancies, inconsistencies remain which indicate that the description of the valence band edge is only a rough approximation to reality. Further discussion of this question is deferred to Section 7.4.

7.2 THE METAL–SEMICONDUCTOR TRANSITION Tl–Tl₂Te

7.2.1   Genesis of the Superposed Band Model

The electronic behavior of Tl_xTe_{1-x} for $x \gtrsim \frac{2}{3}$ has been explained in terms of two bands separated by a small, temperature-sensitive gap, as illustrated in Fig. 7.8. When $x = 1$, on the other hand, one has a typical liquid metal, whose electrical properties are well approximated by the free-electron model. The question of what happens to the electronic structure as the composition is varied between these extremes is an important one. Many other liquid semiconductor alloy systems have an analogous range of behavior with composition, so an examination of the question for Tl–Te may be of broad value.

An experimental study of S and σ over this composition range has been made by Cutler and Petersen (1970) and their experimental curves are shown in Figs. 2.2a and 2.8a. It is seen that σ and S change very rapidly, and with decreasing x there is a continuous progression from metallic to the semi-conductor behavior discussed in the preceding section. Believing at the time that excess Tl is in the form of Tl^{3+} ions when $x \gtrsim \frac{2}{3}$, we have proposed a rigid band model to describe the electronic change. According to this model, the conduction band at $x = \frac{2}{3}$ is generated by the Tl$_2$Te molecules, and as x is increased, the source of this band is gradually changed to Tl^{3+} ions. Thus $N(E)$ remains relatively fixed and E_f increases with x in a continuation of the mechanism discussed in Section 7.1. However, the thallium ions are in fact singly charged (Cutler, 1973). At $x = 1$, the conduction band includes the $(6s)^2$ orbitals of the Tl atoms, but when $x \sim \frac{2}{3}$, the $(6s)^2$ states of the excess Tl^{+1} ions are below the conduction band edge. This is difficult to explain with a rigid band model.

The pseudogap model proposed by Mott (1971b) and outlined in Section 5.1 suggests a way out of the difficulty. According to this model, the effective potential in the electron gas increases on adding Te to Tl, with the result that a minimum develops in $N(E)$ near E_f. The behavior of expanded mercury was the prototype for the pseudogap model, and it is useful to review in more detail the present understanding of what happens when a metal such as Cs or Hg is gradually expanded at supercritical temperatures. This problem has been studied extensively both experimentally and theoretically (Hensel and Franck, 1968; Yonezawa et al., 1974). As shown in Fig. 7.13, σ decreases rapidly as the concentration c of Cs or Hg is decreased below the maximum c_0. At a certain point, which is visible in the curve for Cs, there is a break in the curve with a sudden drop in σ which is identified as a metal-insulator transition. The change in the electronic structure can be best understood in terms of the tight-binding model. At c_0, the bands due to the valence electron states are very broad. The s, p, and higher bands are merged into a free electronlike band. As c decreases, the bands narrow, and ultimately separate. In the case of Hg, which has two valence electrons, E_f lies between the s and p bands, and the metal-insulator transition occurs when a "Wilson gap" opens up. In the case of Cs, with one valence electron, E_f lies in the center of the s band. Band narrowing increases the correlation energy between electrons, and a Mott–Hubbard gap develops at E_f as the result of Coulomb repulsion between pairs of electrons on the same atom (Mott, 1974b).

There is evidence, to be discussed shortly, that Tl_xTe_{1-x} is a pseudobinary mixture of Tl + Tl$_2$Te at all compositions $x > \frac{2}{3}$. This suggests that the electronic structure of the alloy is similar to an expanded metal except that neutral Tl$_2$Te molecules take the place of empty space between the Tl atoms. Since Tl has three valence electrons, it should be similar to Cs except that the Mott–Hubbard gap develops in the p band. It seems reasonable, then, to use this electronic model for an expanded metal as a starting point, and consider the added effects on the electronic structure caused by the presence

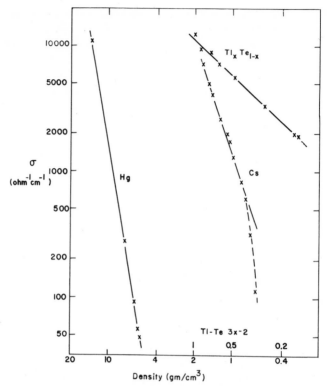

Fig. 7.13. Log-log plots of the electrical conductivity of Hg, Cs, and Tl–Tl$_2$Te as a function of parameters that determine the concentration of the dilute metal (Cutler, 1967a).

of Tl$_2$Te molecules. This approach, taken in later work by this writer (Cutler, 1976a), can be described as a superposed band model, since it consists of adding the density of states curves generated in a tight-binding scheme by the two constituents Tl and Tl$_2$Te. These are assumed to be independent of each other in a first approximation, each containing a number of states proportional to the concentration of the constituent.

This point of view immediately suggests a comparison with metals diluted with other insulator molecules. In recent years, a number of such systems have been studied which are prepared by codeposition from a vapor phase of a metal, such as Na or Cu, and an inert gas such as Ar, on a cold substrate (Cate *et al.*, 1970; Berggren *et al.*, 1974b). Such systems have been found to have a dependence of σ on c which is similar to that of the curves for Hg and Cs in Fig. 7.13. There is also a similar metal-insulator transition. On the other hand, Tl + Tl$_2$Te has a metal-semiconductor transition. In addition, we believe there is evidence of a qualitative difference in the behavior of σ versus c in comparison to Hg and Cs, as shown by the plot for Tl + Tl$_2$Te in Fig. 7.13.

The latter has a slope $-d \ln \sigma/d \ln c = 1$, whereas it is much larger for Hg (5.8) and Cs (3.8).

The problem then is to understand the reason for the qualitative differences in behavior when a metal is diluted in a vacuum or a good insulator and when it is diluted by a substance with a narrow band gap such as Tl$_2$Te. We examine this question in the following subsections. First, we review evidence that the alloy is a mixture of Tl and Tl$_2$Te molecules, and consider the electronic structure of the two constituents. When the tight-binding bands are superposed, the relative energies of the bands of the two constituents are governed by charge transfer between them; in Section 7.2.3 the factors which govern the charge transfer are examined. Next the electronic behavior of Tl + Tl$_2$Te and Tl + Ar are compared in light of the superposed band scheme. This is followed by a qualitative discussion of the distortions from reality which are introduced by the model. Finally we discuss the experimental behavior of S and σ in relation to the model.

7.2.2   The Pseudobinary Alloy Tl$_2$Te + Tl

We have discussed in Sections 3.1 and 3.4 the liquid–liquid phase separation and the thermochemical information which suggest that the alloy is a mixture of Tl$_2$Te + Tl for $x > \frac{2}{3}$. The electrical behavior of Tl alloyed with small amounts of Te indicates that the Te interacts chemically with Tl at low concentrations. The behavior of $\sigma(x)$ at $x \gtrsim 1$ can be analyzed in terms of a scattering cross section A_2 for the Te atoms (see Section 7.2.5), and the resulting value depends strongly on whether the electron density is assumed to remain constant or to be reduced by a chemical reaction between Tl and Te. In the former case, the result is $A_2 \sim 200$ Å^2, which is too large by an order of magnitude. Therefore electrons are removed by trapping (probably in the form of Te^{2-} ions) or by formation of Tl$_2$Te molecules. Flynn and Rigert (1973) have discussed the dependence of the amount of charge trapped (in either real or vitual impurity states) by an electronegative impurity in a metal, and they found that it is a continuous function of the electron density. Their results indicate that Te would not trap any charge for an electron density as high as it is in Tl, which suggests that Tl$_2$Te molecules are responsible for the removal of electrons. This evidence is not conclusive, but we shall proceed on the assumption that there is a mixture of Tl$_2$Te and Tl at all concentrations $x > \frac{2}{3}$. The overall conclusions are expected to be the same if the bonding is ionic, but the mechanics of the charge transfer, which is discussed in the next subsection, would be different.

The tight-binding bands of the two constituents are derived from the energy levels of discrete Tl and Tl$_2$Te molecules, and it is useful to have some idea of the relative energies of the levels closest to E_f. We show in Fig. 7.14 these electronic levels for Tl and Te, taken from atomic calculations (Herman

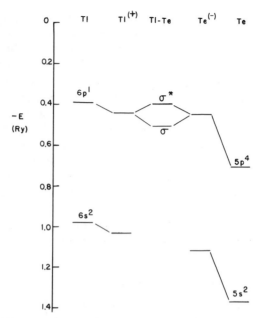

Fig. 7.14.   Energy level diagram for discrete Tl_2Te molecules. A small degree of ionicity is sufficient to shift the Tl(6p) and the Te(5p) orbitals to equal energies under Tl(+) and Te(−1), and the covalent splitting is symmetrical about this energy (Cutler, 1967a).

and Skillman, 1963). For those Tl atoms which bond with Te, the atomic levels are lowered and the Te levels are raised as the result of change transfer in the bond. (The equations for partially ionic bonding are discussed in Section 8.1.) It seems likely that the charge transfer is sufficient to equalize the energies of the bonding orbitals, and the diagram shows the estimated splitting of the σ and σ^* levels due to the Tl–Te bond. (See the original work for details.) The Tl_2Te valence band comes from the nonbonding $(5p)^2$ orbitals of the Te atoms. The conduction band may be derived from the empty 6p orbitals of the Tl atoms as well as the σ^* bond orbitals. Both of these levels are in fact higher than indicated in the diagram since one must add the repulsion energy for bringing another electron into the system. One of the implications of the result in Fig. 7.14 is that $Tl(6s)^2$ and the $Te(5s)^2$ levels are far enough below E_f so that they can be safely ignored in relation to the valence band.

7.2.3 Charging

Let us consider the effect of introducing a void into a metal, and then filling it with Tl_2Te molecule. In Fig. 7.15a we show schematically the potential $V(r)$ due to the Tl ions, and the energies of the band edge and

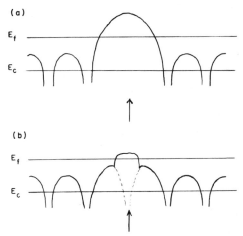

Fig. 7.15.   Schematic representation of the effective microscopic potential of a metal (a) in the presence of a void and (b) in the presence of an insulator molecule. The arrows indicate the position of the impurity (Cutler, 1976a).

the Fermi level. For a large enough void, $V(r)$ will rise above E_f and the electron wave functions are attenuated in the region where $V > E$. If the void is filled with a Tl$_2$Te molecule as indicated in Fig. 7.15b, $V(r)$ is determined by an effective pseudopotential which is lower than the original. Consequently the wave functions at a given E are less attenuated, and in view of tunneling theory, the change is very sensitive to the degree of reduction of $V(r)$. This effect will be much less for a good insulator like Ar, than for a small gap semiconductor like Tl$_2$Te. The decreased attenuation of the wave function reduces the band narrowing which occurs when c is decreased. In addition, there is an enhancement of the charge transfer caused by the spreading of the wave function away from the vicinity of the Tl ions.

Some useful insight about the charge transfer can be obtained from the Thomas–Fermi model for a dilute metal (Cutler, 1969). Each Tl$^+$ ion is at the center of a Wigner–Seitz sphere of radius r_0 such that $c = 3/(4\pi r_0{}^3)$. The Thomas–Fermi equation has boundary conditions V and $dV/dr = 0$ at $r = r_0$, and $V = \infty$ within the Tl$^+$ ion core radius r_c. When c is less than c_0, the spherical shell corresponding to the increased volume is assumed to be filled with homogeneous dielectric medium with dielectric constant K equal to the high frequency value for Tl$_2$Te, as indicated by the diagrams at the top of Fig. 7.16. A value $K \cong 5$ is estimated for both Tl$_2$Te and for the central Tl$^+$ ion. We show in Fig. 7.17 solutions of $N(r) = 4\pi r^2 n(r)$ for several values of r_0 (in atomic units), where $n(r)$ is the electron density. The solid lines are for $K = 1$ and the dashed lines are for $K = 4.6$. The experimental range of r_0 is 4.3 to 9 a.u., so the solution for $r_0 = 6$ a.u. is representative.

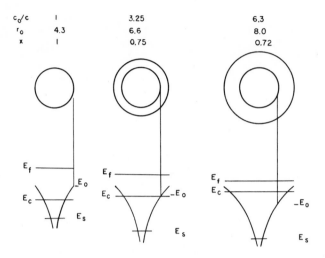

Fig. 7.16. Effect of charging in terms of the Thomas–Fermi model. The circles at the top represent the atomic radius and r_0 (given in atomic units) at various values of c. As c decreases below the value of c_0 for pure Tl, the average potential of the conduction band E_c increases and E_f decreases. The energy E_s of the core states decrease with respect to E_c, as does also the potential E_0 of the Tl_2Te molecules nearest to the Tl^+ ion (Cutler, 1976a).

The effect of the large K is striking. It suggests that a large fraction of the screening charge of the Tl^+ ions is in the region of the Tl_2Te molecules, and the molecules nearest to the ions are in the largely unscreened field of the Tl^+ ions. (The calculation was for a value of $r_c = 1.5$ a.u. which is too small for Tl^+ ions. A larger value of r_c enhances the spreading.)

We show schematically in Fig. 7.16 the effect of the charge transfer on the pertinent energy levels. As c decreases and r_0 increases, the average potential $V(r)$ increases, and this raises the bottom of the conduction band E_c at the same time that E_f is reduced. The increased distance of the screening charge lowers $V(r)$ in the vicinity of the $Tl^+(6s)^2$ electrons, so that their energy E_s is reduced. This, together, with the rise in E_c, enhances the separation of the s and p bands. The effect of the field on the energy of the Tl_2Te states is indicated by the potential energy E_0 of the inner boundary of the Tl_2Te shell. This decreases down as E_c increases, which indicates that the Tl_2Te conduction band moves down with respect to the conduction band of the Tl^+ ions as c decreases.

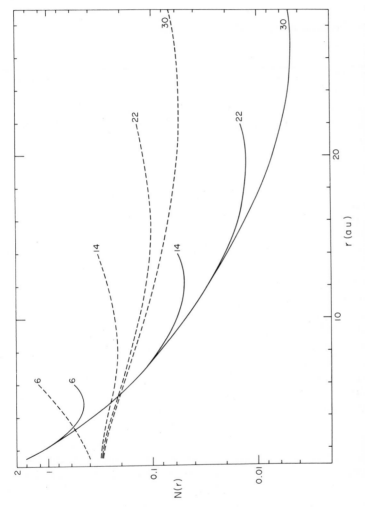

Fig. 7.17. Electron density as a function of radial distance in the Thomas–Fermi model, determined for various values of r_0 (given in a.u.), for the case where $r_c = 1.5$ a.u. The solid curves are for $K = 1$ and the dashed lines are for $K = 4.64$ (Cutler, 1969).

7.2.4 Application of the Superposed Tight-Binding Band Model

Let us first consider the hypothetical alloy Tl_yAr_{1-y}. Argon has a wide band gap, and the Fermi energy in the Tl conduction band must be within this gap. The relative distance of E_f from the argon conduction and valence bands is determined by charge transfer between thallium and Ar atoms, analogous to that discussed in the preceding subsection. This provides the counterpart to the contact potential when there are two separate phases. Since the Ar electrons are held more tightly than those of Tl, the Ar levels are shifted downward and E_f lies closer to the conduction band.

We show schematically in Fig. 7.18a the expected electronic structure as a function of y. For simplicity, only the Tl and Ar conduction bands are indicated, with semicircular curves for the density of states. The only quantitative element is that the area enclosed by $N(E)$ is proportional to the concentration. E_f, indicated by a horizontal line in the Tl band, is expected to remain always below the Ar conduction band. Consequently, the electrical behavior is very similar to expansion in a vacuum. At low y, the Tl conduction band becomes an impurity band within the band gap of Ar, and ultimately a Hubbard gap occurs in the impurity band, causing a metal-insulator transition.

Consider now the behavior of $Tl_y(Tl_2Te)_{1-y}$. We show in Fig. 7.18b the valence band as well as the conduction band of Tl_2Te. Since $E_G \sim 0$ at $y \sim 0$, the centers of the two Tl_2Te bands are separated by the average band width, which we estimate to be 2–3 eV. At $y \sim 1$, the Tl_2Te bands are separated, and E_f is closer to the Tl_2Te conduction band as in Ar. As y decreases, the Tl_2Te bands increase and the band gap decreases in size. Also, charging causes the Tl_2Te bands to shift downwards with respect to E_f. At some point, Tl_2Te conduction band edge drops below E_f, and the Tl electrons partially occupy Tl_2Te conduction band states. For reasons discussed in the next subsection, we think that this occurs when $y \cong \frac{2}{3}$, or $x \cong 0.8$. At smaller y, the occupied part of the conduction band is increasingly derived from Tl_2Te. In comparison to Tl–Ar, the Tl impurity band is absorbed in the conduction band of the semiconductor. This is a well-known phenomenon in highly doped crystalline semiconductors. But in the present case, we have approached it from the other side in composition, which is generally impossible in crystals. At low temperatures it is conceivable that an impurity band separates in Tl–Tl_2Te when $y \to 0$, but we do not think that this happens because of the disorder (Fisher, 1959).

(a)   (b)

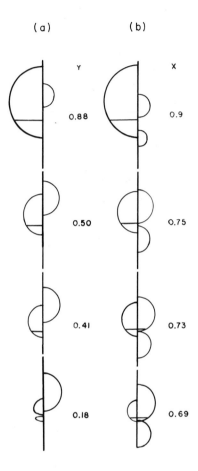

Fig. 7.18. The superposed tight-binding band scheme for (a) Tl_yAr_{1-y} and (b) Tl_y $(Tl_2Te)_{1-y}$. $N(E)$ is shown for several values of y on the left of the vertical line for Tl and on the right for Ar or Tl_2Te. Only the conduction band is shown for Tl and Ar, and the valence band is included for Tl_2Te. E_f, indicated by a horizontal line, is always below the Ar conduction band edge, but it lies above the Tl_2Te conduction band edge when $y < 0.67$ and $x < 0.80$ (Cutler, 1976a).

This model is drastically simplified, and it is well to review some of the factors which are neglected:

(1) The diagrams in Fig. 7.18 do not attempt to depict band narrowing. This depends on the transfer integral between states in a tight-binding band, and it will be strongly affected by the effective potential caused by the other constituent.

(2) Charging causes a greater reduction of energy for Tl_2Te molecules closest to the Tl atoms, and this distorts the shape of the Tl_2Te bands.

(3) There will be a mixing of states between Tl and Tl_2Te bands at nearly equal energies, which will be maximum for equal energies. This will cause a distortion of the shapes of the bands when they approach each other and when they merge. In addition, the states in recognizably separate bands will have some mixed character.

As discussed in Section 5.1, the electronic structure of alloys has long been a subject of intense interest, and the present problem differs from those usually considered in that one of the constituents is a covalently bonded molecule rather than a metal. The Coherent Potential Approximation (CPA) has yielded model calculations which demonstrate the significance of some of the effects neglected in our model, particularly band narrowing and interband mixing. We show in Fig. 7.19 the result of such a calculation from the work of Velicky *et al.* (1968). Note particularly the distortion from the symmetrical shape of the tight binding density of states, and the mixing of parentage of states in the bands. More recent work with the CPA has included consideration of charging effects (Gelatt and Ehrenreich, 1974).

We think that some useful insight can be derived from yet another point of view. The question which we started out with is whether the conduction band of a $Tl–Tl_2Te$ mixture can be approximated by a rigid band model. Such a model implies that the number of states in the band is proportional to the total number of atoms, rather than the number of a particular constituent. Having decided that the rigid band will not do, we went to the other extreme based on tight-binding bands. In that view, the fraction of occupied conduction band states remains constant as c is decreased (until another band overlaps E_f), whereas it decreases in a rigid band model. This is an exaggeration that would be partially remedied if allowance is made for interband mixing, as in the CPA. An approximate guide to the fraction of band states below E_f is provided by the theory of phase shifts of partial waves caused by scattering centers (Friedel, 1954). In this picture, the Tl_2Te molecules act as scattering centers in the electron gas of the Tl atoms, as illustrated by the potential diagram in Fig. 7.15. If the

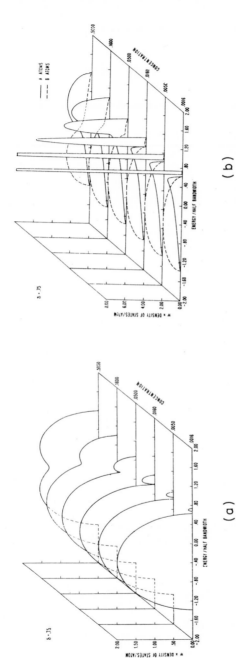

Fig. 7.19. (a) Total density of states and (b) component density of states (per atom of the component) for a model random binary alloy calculated by the coherent potential approximation for the case where $E_A - E_B = 0.75\Gamma$. (Γ is the half bandwidth.) The solid lines in (b) represent the A component, whose concentration is given in terms of the atomic fraction, and the dashed lines represent the B component. (Velicky *et al.*, 1968.)

density of scattering centers is small, the density of states at a given energy E is changed by:

$$\Delta N(E) = \frac{2}{\pi} \sum_l (2l + 1) \frac{d\eta_l}{dE}, \qquad (7.11)$$

where $\eta_l(E)$ is the phase shift of a scattered wave with angular momentum quantum number l.

To understand the effect of introducing an impurity such as Tl_2Te or Ar on $N(E)$, let us approximate the perturbation by a square potential of height V_0. If V_0 is infinite, the wave functions would not penetrate the scattering centers at all, and $\eta_l(E)$ would have the maximum value. But in this limit $-\Delta N(E)$ would be exactly proportional to the volume occupied by the impurity. This corresponds to the approximation of the tight-binding band model. When V_0 is less than infinite, the penetration increases, and $\eta_l(E)$ decreases. When $V_0 \ll E$, $\eta_l \sim 0$ and $\Delta N(E) \sim 0$. This corresponds to the rigid band limit. It is clear then that in reality $-\Delta N(E)$ is intermediate in value, and it becomes large or small depending on the ratio V_0/E. We expect it to be large at $E < E_f$ for an impurity like Ar, but comparable to one for an impurity like Tl_2Te. Of course, Eq. 7.11 is invalid because of multiple scattering except in the limit $y \to 1$, but the qualitative implications can be expected to hold true.

7.2.5 Experimental Behavior

Over most of the composition range, σ is nearly independent of T and decreases rapidly with decreasing x (see Fig. 2.2a). The dependence of σ on x has been analyzed in terms of the electron gas formula:

$$\sigma = e^2 n^{2/3} \lambda_s / \hbar (3\pi^2)^{1/3}, \qquad (7.12)$$

where n is the electron density and λ_s is the scattering distance (Cutler and Petersen, 1970). Assuming that n is determined by the concentration of unbound Tl, λ_s was calculated versus composition, and analyzed in terms of scattering contributions from Tl(1) and $Tl_2Te(2)$ according to the formula:

$$1/\lambda_s = \phi_1/\lambda_1 + \phi_2/\lambda_2, \qquad (7.13)$$

where ϕ refers to the volume fraction. Schaich and Ashcroft (1970) have given a theoretical justification for this formula. The scattering distance λ_2 due to Tl_2Te is inversely proportional to its concentration, and it is conveniently written in terms of the scattering cross-section A_2. It was found that λ_s varies linearly with the Tl_2Te concentration as suggested by Eq. 7.13 with $A_2 = 23.3 \text{ Å}^2$. This compares with the geometric size 26 Å^2 of Tl_2Te (assuming a spherical shape). If the electron gas has one electron per thallium atom instead of three as assumed in the original work, A_2 has the value 11.3 Å^2, which suggests a "softer" scattering center. As x decreases, the second term in Eq. 7.13 rapidly becomes dominant, and the fact that σ becomes independent of T agrees with the expectation that A_2 is independent of T.

The magnitude of S and its dependence on T for pure Tl is in accord with the Ziman theory (Faber, 1972). When Te is added, S becomes proportional to T, as expected from the metallic formula (Eq. 6.4b) when λ_s is independent of T. This behavior persists to $x \sim 0.8$, but for lower values of x, $-S$ rises less rapidly than T. Another change which occurs at this composition is visible in the plot of σ versus S, shown in Fig. 7.20. On a log–log plot, there are two straight line portions with a break at $x \sim 0.8$. At lower σ the slope is -1, in accord with the rigid band model derived in Section 7.1. This break reflects an abrupt change in S versus x rather than σ versus x, which is smooth (see Fig. 7.13). We think that the irregularity indicates that the Tl₂Te conduction band edge crosses E_f at $x \sim 0.8$. Referring to Eqs. 6.4, S depends on $d\sigma/dE$ at E_f, and hence is more sensitive than σ to a change in shape of $\sigma(E)$ near E_f.

This conclusion also resolves the long-standing problem of explaining the dependence of S on T in the range $0.7 < x < 0.8$ (Cutler and Petersen, 1970). The decrease in $-S/T$ with T suggests the presence of a valence band not far below E_f. The rigid overlapping band model discussed in Section 7.1 succeeds in explaining the dependence of S on T for $x < 0.7$, but at higher

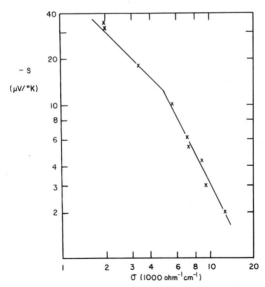

Fig. 7.20.   Plot of $\ln(-S)$ versus $\ln \sigma$ for Tl$_x$Te$_{1-x}$ at $T = 900°$K in the range $x > \frac{2}{3}$. The change in slope occurs at $x \cong 0.80$ (Cutler, 1976a).

values of x the predicted value of $-dS/dT$ becomes too large (Cutler, 1974a). The reason is that E_f is too far above the valence band edge for holes to make an appreicable contribution to S. But the superposed band model indicates that the conduction band cannot be rigid in this range, but contains two parts due to Tl and to Tl_2Te. The latter moves upward with respect to E_f as x is increased, and the valence band follows it as shown in Fig. 7.18b. This provides a natural explanation of the persistence of electron–hole excitations which are visible in $S(T)$ up to $x \cong 0.8$.

7.3 S-TYPE RANGE: $x < \frac{2}{3}$

The first effort to derive a quantative explanation of the behavior of the Te–rich alloys in terms of bond breaking (Cutler, 1971a) was unduly complicated because a model of the band structure had to be formulated at the same time that a theory for bond equilibrium was examined. Later work (Cutler, 1976b) has made it possible to deduce the electronic structure of the valence band directly from the experimental data for σ and S, in terms of a rigid band model. This result, presented in the first subsection, simplifies the overall problem, and makes it possible to compare the experimental behavior of the hole concentration directly with the predictions of bond equilibrium theory. The latter theory was first developed in a relatively complicated form (Cutler, 1971b). We discuss an improved version in Section 7.3.2 (Cutler, 1976d), and in the final subsection it is compared with the experimental behavior.

7.3.1 Rigid Band Model for the Valence Band

A test of whether rigid band behavior occurs, as discussed in Section 7.1, has been to show that a plot of an appropriate function of S and σ causes the experimental points to fall on a single curve, from which the form of $\sigma(E)$ can be deduced. In Section 7.1, it was tacitly assumed that $N(E)$ is invariant with x. This is reasonable for a small range of composition ($\frac{2}{3} < x < 0.7$). In the present case, a large range of composition is pertinent, and rigid band behavior is a less obvious possibility. When this question was first examined (Cutler, 1971a), an incorrect conclusion that $N(E)$ depends on x resulted from a failure to eliminate all data affected by ambipolar transport, which can also cause points to deviate from a common curve. The information derived in Section 7.1.2 indicates that the band gap is negative when $T > 770°K$, so that ambipolar transport may occur at higher T unless E_f is well below the valence band edge. Experimental curves for $\ln \sigma$ versus T^{-1} for various compositions of Tl_xTe_{1-x} in the range $x < \frac{2}{3}$ are shown in Fig. 2.2b. The curves are nearly parallel, and characteristic deviations from the common shape occur at high T and $x > 0.6$ which indicate ambipolar effects. The thermopower is more sensitive to ambipolar effects, but the curves for $S(T)$, shown in Fig. 2.8b, fail to display a pattern which distinguishes an ambipolar region. Consequently, a somewhat arbitrary prescription was used. Data were eliminated as suspect if $T > 770°K$ and $x > 0.4$. When this is done, rigid band behavior becomes evident, as is shown below.

In testing for rigid band behavior, the need to guess in advance the shape of $\sigma(E)$ can be avoided if the metallic approximation is used. Then Eqs. 6.4a and 6.4b show that a plot of S/T versus σ should yield points on a common curve. We show such a plot in Fig. 7.21, with suspected ambipolar points eliminated. The points fall on a common curve for compositions $0.2 \leq x \leq 0.6$, except at low σ where the metallic approximation becomes poor. The data for pure Te lies on a separate curve, which indicates that the rigid band model fails when $x < 0.2$. In the middle range ($\sigma \sim 500$ ohm^{-1} cm^{-1}) $d \ln(S/T)/d \ln \sigma = -1$, which corresponds to $\sigma(E) \propto E$. This form of $\sigma(E)$ accounts for the behavior at $\sigma \gtrsim 700$ ohm^{-1} cm^{-1}, as can be seen when σ and S are expressed in terms of Fermi–Dirac integrals. Equations 7.1 to 7.6 provide the appropriate description if we let f, E and E_f refer to hole energies, change the sign of e in Eq. 7.5, replace n by p and E_{c0} by E_{v0}. In Fig. 7.22, the experimental plot of $\ln(\sigma/T)$ versus S is compared to

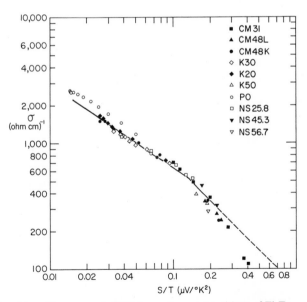

Fig. 7.21. Plot of $\ln \sigma$ versus $\ln(S/T)$ for various compositions of Tl_xTe_{1-x} in the range $x < \frac{2}{3}$. The numbers indicate the compositions in at% Tl (from Cutler, 1976b).

the theoretical curve with $A_p C_p^2 = 2960$ ohm^{-1} cm^{-1} eV^{-1}, which corresponds to the constant value of $S\sigma/T$ in Fig. 7.21 near $\sigma \sim 500$ ohm^{-1} cm^{-1}. The uncircled points fall on the theoretical curve marked $x_c = 0$. The circled points, which are in the range where ambipolar effects are suspected, show the decrease in S caused by electron transport.

As mentioned in Section 7.1.1, the effect of a mobility edge at E_{v1} can be examined with the help of modified Fermi–Dirac integrals discussed in Appendix B3. They are introduced by replacing Eqs. 7.4 and 7.5 by Eqs. B8a and B8b. Since the experimental range of T is relatively small, $x_c = (E_{v1} - E_{v0})/kT$ is roughly constant and it is reasonable to make comparison with the theoretical mobility shoulder curves for constant x_c. These curves are shown in Fig. 7.22 for $x_c = 1$ and 2. The $x_c = 2$ curve fits poorly, but the one for $x_c = 1$ is somewhat better than $x_c = 0$. This indicates that there

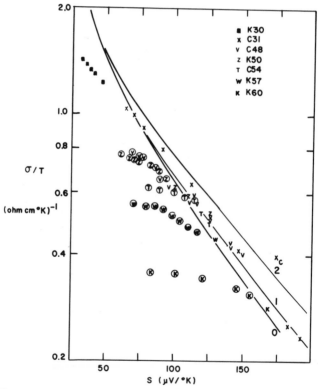

Fig. 7.22. Plot of $\ln(\sigma/T)$ versus S for various compositions of $Tl_x Te_{1-x}$ where $x < \frac{2}{3}$. The compositions are given in at% Tl. The circled data points are for $x > 0.4$ and $T > 770°$K. The theoretical curves are for $A_p C_p^2 = 2960$ ohm^{-1} cm^{-1} eV^{-1} with different values of the mobility edge parameter x_c (from Cutler, 1976b).

may be a mobility edge at a distance $kT \sim 0.06$ eV from E_{v0}, but difference in fit for the two curves is probably not significant enough to prove the existence of an edge. The value $x_c = 1$ corresponds to $\sigma(E_{v1}) = 200$ ohm^{-1} cm^{-1}, which agrees with Mott's estimate (1972a).

We turn now to the behavior at large σ. The change in slope at $\sigma \gtrsim 700$ ohm^{-1} cm^{-1} in Fig. 7.21 implies that $\sigma(E)$ increases less rapidly than E at larger E. The metallic approximation is good in this range, and one can integrate directly the empirical function $dE/d \ln \sigma$ which is obtained from Eqs. 6.4. This is conveniently done with the help of a straight line approximation to the experimental curve shown in the figure. It corresponds to $\sigma(S/T)^{0.645} = 151$, where the units in the figure are used. Using $\sigma = 2960E$ when $\sigma = 700$ ohm^{-1} cm^{-1} as the boundary condition for integration, the overall result is:

$$\sigma = 2180(E - 0.065)^{0.643} \quad \text{ohm}^{-1} \text{ cm}^{-1} \quad \text{for} \quad E > 0.235 \text{ eV,}$$
$$\sigma = 2960E \qquad\qquad\qquad\qquad\qquad \text{for} \quad E < 0.235 \text{ eV.} \qquad (7.14)$$

This result can be used to determine E_f as a function of T and x from experimental values of σ, except in the relatively small range $x > 0.6$ where ambipolar effects occur, and of course $x < 0.2$ is also excluded. In the metallic range ($\sigma \gtrsim 500$ ohm^{-1} cm^{-1}) $\sigma = \sigma(E_f)$ and Eq. 7.14 is simple inverted. In the Fermi–Dirac range, it is convenient to use an analytical expression which exists for $F_0(\xi)$, so that Eqs. 7.4 and 7.14 give:

$$E_f - E_{v0} = kT \ln[\exp(\sigma/2960kT) - 1]. \qquad (7.15)$$

The quantity $N(E)$ is obtained from Eqs. 6.14 and 7.14 with an unknown constant A_p. Knowing $E_f(x, T)$, as described above, Eq. 7.3 can be used to determine how p varies with x and T again with an unknown constant. In order to have convenient units, we calculate $c^* = p^*/N_a$, where $N_a = 2.7 \times 10^{22}$ cm^{-3} is the concentration of atoms. The quantity p^* is the value obtained if A_p has the value $A_1 (= 8.86 \times 10^{-42}$ eV^2 cm^6 ohm^{-1} cm^{-1}), which corresponds to $\lambda = 1$ and $a = 3$ Å in Eq. 6.14b. The metallic approximation gives the result:

$$c_M{}^* = 0.0091 + (\sigma/3250)^{2.06} \quad \text{for} \quad \sigma > 700 \quad \text{ohm}^{-1} \text{cm}^{-1}$$
$$= (\sigma/5030)^{1.5} \quad \text{for} \quad \sigma < 700 \quad \text{ohm}^{-1} \text{cm}^{-1}. \qquad (7.16)$$

In the lower range of σ, Fermi–Dirac integrals (Eq. 7.3) are more accurate, and the result $c_F{}^*$ cannot be expressed analytically. It is convenient to compute composite curves for c^* versus x and T by using $c_M{}^*$ for $\sigma > 900$ ohm^{-1} cm^{-1} and c_F otherwise. The result, labeled c_{MF}^*, was calculated from experimental values of $\sigma(x, T)$ and is shown in Fig. 7.23. It is seen that plots of $\ln c_{MF}^*$ versus T^{-1} reflect the nearly parallel behavior of the $\ln \sigma$ curves

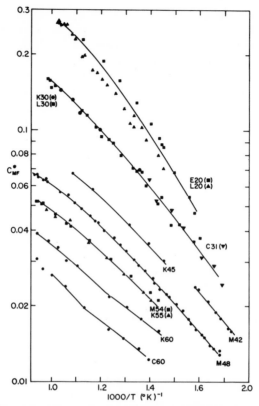

Fig. 7.23. Plots of ln c_{MF}^* versus $1/T$ for various compositions of $Tl_x Te_{1-x}$ (given in at% Tl) as derived from $\sigma(T)$ data (from Cutler, 1976b).

in Fig. 2.2b. The curves have a nearly constant slope which tends to decrease in magnitude at high T. This dependence on T is one aspect of the experimental behavior which needs to be explained by the bond equilibrium theory or other theory for hole formation. The dependence of c^* on x at constant T needs also to be explained. We show in Fig. 7.24 experimental points for σ versus x taken from various sources at $T = 800°K$, together with the derived curve for c_{MF}^* versus x.

The true value of c ($= p/N_a$) is related to c^* by $c = c^*(A_p/A_1)^{1/2} = c^*(a/3 \text{ Å})/\lambda$. The quantity $a = 4.1$ Å, and it seems likely that $0.1 < \lambda < 1$. Therefore $\lambda = 0.3$ yields an estimate $c = 2.8c^*$ which should be accurate within a factor 3. With this conversion, Eq. 7.16 can conveniently be used to estimate the hole concentration for the value of σ for all p-type Tl–Te alloys.

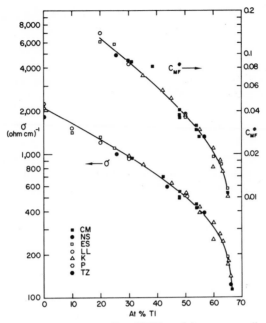

Fig. 7.24. Isotherm of σ versus x at $T = 800°K$, and the corresponding values of $c^*_{MF}(x)$ (from Cutler, 1976b).

7.3.2   Broken Bond Theory for Hole Generation

In Appendix C it is shown that a very simple model for bond equilibrium (the independent bond model) yields the result:

$$\kappa = \frac{c^2}{1 - 3x/2 - c/2},\qquad(7.17)$$

where c is the concentration of dangling bonds normalized to the density of atoms. The equilibrium constant κ is related to the energy E_d and the entropy S_d of formation of a dangling bond by:

$$\kappa = 2 \exp[-2(E_d - TS_d)/kT].\qquad(7.18)$$

(No distinction is made between energy and enthalpy in our discussions.) If the dangling bonds form fully ionized acceptor states, as we believe to be the case, c will also be equal to p/N_a. However in that case, E_d may not be a constant, since it will include the energy of formation of a Te^- ion and a hole. Therefore, before making a comparison of bond equilibrium theory with experiment, it is desirable to discuss the electronic configuration of acceptor ions, and their energy of formation.

In Section 5.3 we described how a dangling bond state is formed when a Te–Te bond is broken. According to Fig. 5.4a, an energy $E_1 = E_\pi - E_\sigma$ is required to form a dangling bond in an isolated molecule. It is necessary now to examine what happens when this occurs in a condensed phase, and a hole is formed in the valence band. The dangling bond forms an acceptor state which can receive an electron from the band, to produce a Te⁻ ion and a hole. The absence of the bond partner creates an effective negative charge when the dangling bond state is doubly occupied, which perturbs the band, as indicated in Fig. 7.25a. If the concentration of mobile screening charge is small, a bound state is created for a hole, i.e., an acceptor state above the valence band edge. This is a well studied phenomenon in crystalline semiconductors, and the energy of formation of the acceptor ions $E_a - E_{v0}$ has

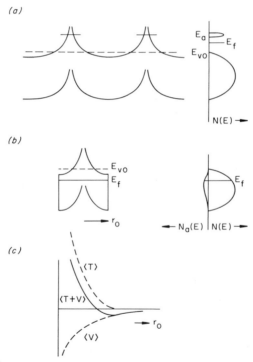

Fig. 7.25. Electronic behavior of acceptor impurity states due to dangling bonds. (a) Low hole densities. The spatial dependence of the band potential is shown on the left and the resulting density of states is on the right. Discrete acceptor states are formed above the band edge E_{v0} (dashed line). (b) High hole densities. The charge and potential distribution is approximated by the Thomas–Fermi model, illustrated on the left, and the impurity state is a virtual level in the band, illustrated on the right. (c) Qualitative behavior of the kinetic energy $\langle T \rangle$ and potential energy $\langle V \rangle$ and their sum as a function of r_0.

been calculated with good accuracy with the use of Debye–Hückel theory for screening (Harvey, 1961).

At higher acceptor concentrations, $E_a - E_{v0}$ decreases both because of increased overlap of the ion potentials and increased screening. Complicated phenomena occur at higher acceptor concentrations, such as formation of an impurity band and merging with the valence band, but these things need not concern us here. We focus attention on the energy of the acceptor states in the case where $E_a \gtrsim E_{v0}$. The acceptor states are then virtual localized states in the valence band as illustrated in Fig. 7.25b. We think that the width of the virtual impurity states may be comparable to the valence band width because of the electronic similarity of the impurity. In this situation, E_f will be below E_{v0}, Fermi–Dirac statistics are applicable, and the kinetic energy plays a more conspicuous role in determining the value of $E_a - E_{v0}$. The electronic configuration is represented by a Wigner–Seitz (WS) spherical volume about each acceptor ion with radius $r_0 = (3/4\pi p)^{1/3}$ which is electrically neutral, as shown in Fig. 7.25b. This is the same as the problem of the charge distribution about a Tl^+ ion, discussed in Section 7.2.2, and the diagram is the same as the ones in Fig. 7.16, but turned upside down. The Thomas–Fermi (TF) model discussed there, can be used in the present case, but now it is necessary to determine the sum of the average potential energy $\langle V \rangle$ and the average kinetic energy $\langle T \rangle$. Because of the inversion from electrons to holes:

$$E_a - E_{v0} = -\langle T \rangle - \langle V \rangle. \tag{7.19}$$

Before examining the Thomas–Fermi solution, it is helpful to consider a simpler model. The quantity $\langle T \rangle$ is taken to be simply the kinetic energy of a free electron gas with density p, and $\langle V \rangle$ is calculated on the assumption that the hole density is uniform within the WS sphere and there is a dielectric constant K. This gives:

$$\langle T \rangle + \langle V \rangle = \frac{3h^2}{10m} \left(\frac{9\pi}{4} \right)^{2/3} \cdot \frac{1}{r_0^2} - \frac{3}{10K} \frac{e^2}{r_0}. \tag{7.20}$$

This expression shows that $\langle T + V \rangle$ is negative and small in magnitude at large r_0, being dominated by the potential energy term. As r_0 decreases, $\langle T \rangle$ becomes increasingly important, causing the sum to go through a minimum and then increase rapidly, as shown in Fig. 7.25c.

The hole density is very much concentrated near the Te^- ion at large r_0, and the TF model gives a much better approximation to $\langle T + V \rangle$. Using calculations from the original work (Cutler, 1969), this was calculated for $K = 1$ and $K = 3$, as shown in Fig. 7.26. An abscissa scale in terms of σ is also shown, assuming $c/c^* = 5$ (see Eq. 7.16). It is seen that in the case of $K = 1$ the magnitude of $\langle T + V \rangle$ is less than $kT(\sim 0.06\,\text{eV})$ for $r_0 > 8$ a.u. corresponding to $\sigma \sim 400\,\text{ohm}^{-1}\,\text{cm}^{-1}$, and it increases rapidly at smaller

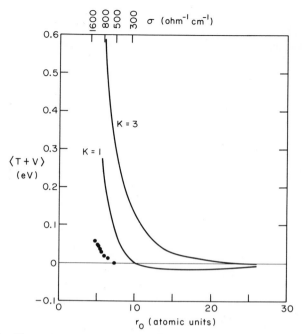

Fig. 7.26. Thomas–Fermi solutions for $\langle T + V \rangle$ versus r_0 for $K = 1$ and $K = 3$. The conductivity scale is based on the assumption that $c = 5c^*$. The points indicate the values of $-\Delta E_d$ required to reconcile the IBM and experimental curves for $c(x)$ in Fig. 7.28.

r_0. For $K = 3$, $\langle T + V \rangle$ changes appreciably over the whole experimental range. The TF model is still a rather crude model. Aside from the neglect of correlation and exchange, it treats the dielectric screening in a very simplistic way in a range where complicated changes occur in that phenomenon. When E_f is well below the band edge, the effective value of $K \sim 1$, but it may be larger when E_f is close to E_{v0}.

The result in Fig. 7.26 suggests that $E_{v0} - E_a$ can be expected to be small and relatively constant at low values of p. As p increases, E_a is expected to start to decrease appreciably at some point. Since $E_d = E_1 + E_a - E_{v0}$ appears in an exponential (Eq. 7.18), κ will increase rapidly when $\Delta E_a \gtrsim kT$.

7.3.3 Comparison with Experiment

Equation 7.17 describes the result of an unrestricted independent bond model (IBM) which assumes that the change in free energy for breaking a Te–Te bond is independent of whatever else may be attached to the two atoms. In Fig. 7.27 the IBM curves for $\ln c$ are plotted versus $\theta = (E_d - TS_d)/kT$ at various values of x, and it is seen that they are qualitatively similar to the

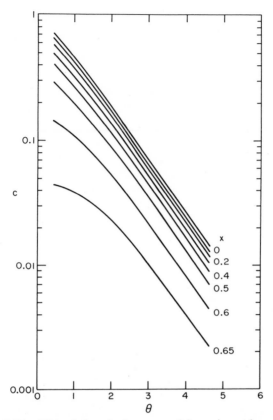

Fig. 7.27.   IBM solutions for $\ln c$ versus θ for various values of x.

experimental curves in Fig. 7.23. The constant slope at large θ reflects the situation where $c \ll (1 - 3x/2)$ in Eq. 7.17, and the slight curvature at smaller θ reflects incipient saturation in the broken bond density ($c = 2 - 3x$) which occurs at large negative values of θ. The shape of the most precise curve for $c^*(T)$ (M48 in Fig. 7.23) can be reproduced within experimental error with the values $E_d = 0.234$ eV and $S_d = 2.0\,k$, and matching of the ordinates gives $c/c_{MF}^* = 4.28$. We show in Fig. 7.28 the IBM isothermal curve $c(x)$ for $\theta = 1.39$, corresponding to $T = 800°$K, together with the superposed experimental curve from Fig. 7.24 with $c = 3.9\,c_{MF}^*$. The fit is reasonably good for $x > 0.40$, but the theoretical curve increases too slowly at lower x. It is too low by a factor of two at $x = 0.2$.

   The IBM is a very simplified model for bond equilibrium, and one may ask whether the discrepancy in Fig. 7.28 can be decreased by a reasonable modification of the model. One possible refinement is to allow E_d and S_d to be

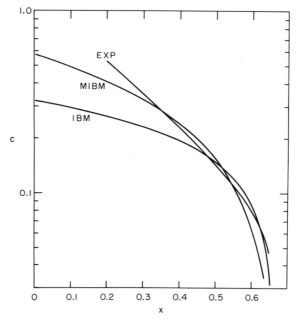

Fig. 7.28. Comparison of theoretical curves (IBM and MIBM) for $c(x)$ with the experimental curve from Fig. 7.24.

changed by proximity of the dangling bond to a Tl atom. Equations are derived in Appendix C.1 for a modified theory (MIBM) in which the free energy of a dangling bond is altered if the Te atom is attached to a Tl atom. The parameter β in Eq. C18 is given by:

$$\beta = (\kappa/2)^{1/2} \exp[(E_{d1} - TS_{d1})/kT], \qquad (7.21)$$

where E_{d1} and S_{d1} are the parameters for the Tl–Te* dangling bond. If $\beta \gg 1$ over the entire temperature range, the effect is essentially to reduce $c(T)$ by a factor which is nearly constant, but which decreases as x decreases. It is possible to match the experimental dependence of c^* on T^{-1} with a wide range of values of E_{d1} and S_{d1}, and yet modify the predicted isothermal curve $c(x)$. We show one such result in Fig. 7.28. It is seen that it was possible to get a better overall match to the experimental curve at the price of somewhat larger deviations at $x \gtrsim 0.6$. It should be mentioned that the MIBM yields different values of E_d, S_d, and c/c^*, depending on the choice of E_{d1} and S_{d1}. In the example shown, $c/c^* = 4.9$, $E_d = 0.234$ eV, $E_{d1} = 0.467$ eV, $S_d = 2.94$ k, and $S_{d1} = -5.9$ k.

Another possible explanation for the IBM discrepancy at low x is that E_d may be decreasing due to the increasing value of c, as discussed in Section

7.3.2. It is a simple matter to calculate the change in θ required to match the IBM curve to the experimental curve in Fig. 7.28, and deduce from this the change ΔE_d as a function of c. We have plotted $-\Delta E_d$ in Fig. 7.26, and it is seen that the rise in $-\Delta E_d$ at low c is abrupt as predicted by the TF model, but it occurs at a larger c. However, this explanation implies a drastic change in the dependence of c on T^{-1}, and values of $-\Delta E_d > kT$ cause the theoretical curves for $x = 0.2$ to rise much more rapidly at high T than what is observed experimentally. However, one may observe that the experimental curves for $x = 0.2$ and 0.3 in Fig. 7.23 have perceptively larger slopes than those for larger x, in contrast to the theoretical curves in Fig. 7.27. This suggests that a more modest increase in $-\Delta E_d$ ($<kT$) is in fact occurring, of a magnitude suggested by the smaller difference between the MIBM and experimental curves in Fig. 7.28. When ΔE_d calculated by the MIBM comparison was used to calculate $\ln c$ versus T^{-1} for $x = 0.2$ and 0.3, there was in fact better agreement than the IBM with the experimental curve.

The IBM model also yields predictions about the thermochemical behavior. It is convenient to use pseudobinary mixing functions which are defined in the same manner as Eq. 3.1. For the mixture Te + Tl_2Te:

$$\Delta G_m{}^*(x) = \Delta G(x) - \left(1 - \frac{3x}{2}\right)\Delta G(0) - \left(\frac{3x}{2}\right)\Delta G\left(\frac{2}{3}\right). \qquad (7.22)$$

The pseudobinary entropy of mixing $\Delta S_m{}^*$ and enthalpy of mixing $\Delta H_m{}^*$ are defined in the same way. $\Delta G_m{}^*$ obeys an equation similar to Eq. 3.3:

$$\Delta G_m{}^*(x) = \left(1 - \frac{3x}{2}\right)\left[\mu_{Te}(x) - \mu_{Te}(0)\right] + \frac{x}{2}\left[\mu_{Tl_2Te}(x) - \mu_{Tl_2Te}\left(\frac{2}{3}\right)\right]. \qquad (7.23)$$

The differences in chemical potential are related to concentration parameters of the IBM theory which is derived in Appendix C1:

$$\mu_{Te}(x) - \mu_{Te}(0) = RT \ln(y/y_0), \qquad (7.24)$$

$$\mu_{Tl_2Te}(x) - \mu_{Tl_2Te}(\tfrac{2}{3}) = RT \ln(L/L_0), \qquad (7.25)$$

where y is given by Eq. C16 and L is given by Eq. C17. The quantity y_0 is evaluated at $x = 0$ and L_0 is evaluated at $x = \frac{2}{3}$. Using the above equations, one can caluclate $\Delta G_m{}^*(x)$. The quantity $\Delta S_m{}^*$ is obtained by the thermodynamic relation $\Delta S_m{}^* = -\partial\, \Delta G_m{}^*/\partial T$, and $\Delta H_m{}^* = \Delta G_m{}^* + T\, \Delta S_m{}^*$.

In the top part of Fig. 7.29 we compare $\Delta G_m{}^*/RT$ with the experimental curve (Nakamura and Shimoji, 1971). It is seen that the IBM curve has the right general shape, but is $\sim 2x$ too small in magnitude. Below, $\Delta S_m{}^*$ is compared with both the experimental curve and the theoretical curve for ideal mixing (Eq. 3.6). It is seen that the agreement with experiment is much better here, and the error is comparable to that of the ideal mixing formula. Insight

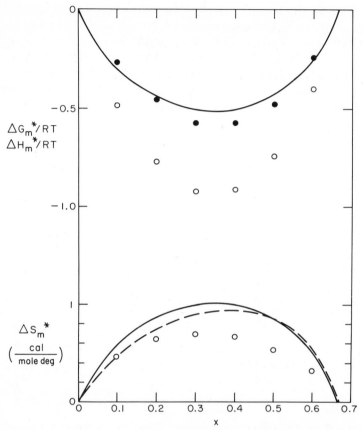

Fig. 7.29. Comparison of IBM theory (solid line) and experiment (open circles) for ΔG_m^* and ΔS_m^* for the pseudobinary mixture $Tl_2Te + Te$ at 600°C. The closed circles are experimental data for ΔH_m^*. The dashed line is theoretical curve for the ideal entropy of mixing from Eq. 3.6.

into the reason for this behavior is provided by comparison of ΔH_m^* with experiment. The experimental curve for $\Delta H_m^*/RT$ is shown in the top part of the figure. It has a minimum value ~ 0.6, and a shape in agreement with the expected behavior for the heat of mixing of Tl_2Te with tellurium (Eq. 3.5). In contrast, the theoretical values of ΔH_m^* are too small to be seen on the plot. It is easy to see that the reason for this is built into the assumptions of the IBM. It is assumed that the only energy of interaction occurs through the mechanism of creating dangling bonds, and these are in a very low concentration. The neglected secondary interactions between molecules cause an enthalpy of interaction of the order RT. The lack of this ingredient in the

IBM accounts also for the discrepancy in $\Delta G_m{}^*$. It is probably also sufficient to account for the smaller deviation in $\Delta S_m{}^*$, since the secondary interactions should also make a contribution to the entropy of mixing. But a large fraction of $\Delta S_m{}^*$ comes from the configurational entropy, and the IBM and Flory formula (Eq. 3.6) yield somewhat different results for this.

The neglected secondary bonding effects could affect E_d and S_d, but only to the extent that they differ for Te atoms with intact or dangling bonds. Such a contribution ΔS_d would change the shape of the isothermal $c(x)$ curve without changing the temperature curves, but secondary bonding contributions ΔE_d would change both $c(T)$ and $c(x)$ as noted in the earlier discussion.

In summary, the IBM model seems to provide a reasonable explanation for the observed dependence of p on x and T, if allowance is made for the extreme simplifications which it contains. The deviations from experiment which are observed have a number of possible explanations, and because of this, it is not possible to pin down which of them it is, or whether the truth lies in another direction.

7.4   THE PSEUDOGAP

So far, models have been used and largely justified in which the conduction and valence band edges are parabolic. However there are indications of band tailing, particularly in the valence band. The overlapping band model (Section 7.1) yields curves for $\sigma(T)$ (Fig. 7.9) which are somewhat flatter than the experimental curves at $T < 770^\circ K$, where $E_G > 0$; this suggests that a valence band tail is already overlapping the conduction band edge. Another indication is the fact that different values of $AC_p{}^2$ (denoted here by B) have been deduced from analyses of bodies of data corresponding to varying ranges of value of E_f. The overlapping band analysis, for which $E_f > E_{v0}$, yielded a value, $B_1 = 422$ ohm^{-1} cm^{-1} eV^{-1}. In the original work (Cutler, 1974a), an analysis was also made of the ambipolar contributions to $\sigma(T)$ for $0.60 < x < 0.66$ which led to a larger value $B_2 = 835$ ohm^{-1} cm^{-1} eV^{-1}. In this case, $E_f \cong E_{v0}$. Finally, as discussed in Section 7.3.1, analysis of the data for $x < 0.60$, where $E_f < E_{v0}$, yielded a value of $B_3 = 2960$ ohm^{-1} cm^{-1} eV^{-1}. Only in the last case is there internal evidence that justifies the parabolic shape. This suggests that there is a tail above the valence band edge as illustrated in Fig. 7.30. In situations where $E_f \gtrsim E_{v0}$, where E_{v0} is the "true" parabolic edge corresponding to B_3 as shown in the figure, models which assume a parabolic shape would result in a projection of the actual shape into the parameters of the model. As indicated by the dashed curves, N_{v1} and N_{v2}, corresponding to Fermi energies at E_{f1} and E_{f2}, such models lead to smaller apparent values of C_p and higher

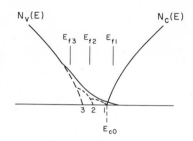

Fig. 7.30. Effect of a valence band tail on the apparent form of the valence band edge (dashed curves) if a parabolic shape is assumed in analyzing data at various values of E_f shown in the figure.

apparent positions of the valence band edge. This evidence of a tail in the valence band suggests that the true band gap is larger than the value $E_G^{(1)} = 0.58 - 7.5 \times 10^{-4}T$ eV deduced in Section 7.1.

As a device for making a better estimate of the band gap, we have used Eq. 7.15 to calculate $E_f^*(T)$ from $\sigma(T)$ in the composition range $x > 0.60$, where ambipolar effects and band overlap cause deviations from the extrinsic behavior (Fig. 2.2b). The quantity E_f^* differs from the true Fermi energy, of course, when $E_{c0} \gtrsim E_{v0}$, but we are looking for an increase in slope of E_f^* versus T which is evidence of band overlap. The results are shown in Fig. 7.31. Two regions are evident. At lower T and x, the curves have a lower slope which corresponds to the normal extrinsic dependence of E_f on T when the band gap is large. At higher T, the curves bend up and merge into an intrinsic common curve which reflects the overlapping band situation. In this range, E_f^* differs from E_f, of course, but the true behavior of E_f is likely to be not much different. Curve 1 marks the value of $E_{v0} - E_{c0}$ derived from $E_G^{(1)}(T)$, and it is seen that it is significantly higher than the range of values $E_{v0} - E_f^*$ in the overlapping band region. Curve 2 is drawn with the same slope as curve 1, so as to overlap the intrinsic curve at the higher temperatures. This yields an estimate $E_G^{(2)}$ which is larger than $E_G^{(1)}$ by 0.12 eV. The smaller slope of the intrinsic curve has two likely causes. At lower T, $E_{v0} - E_f < kT$, and statistical effects cause deviations from the metallic model which we are implicitly using. For $E_{v0} - E_f \gtrsim 0.08$ eV, the metallic approximation is not so bad, but as noted earlier, E_f^* differs from E_f when the bands overlap. Using a simple two-band metallic model (and ignoring the overlap term such as occurs in Eq. 7.10), the expected behavior of $E_{v0} - E_f^*$ assuming $E_{c0} - E_{v0} = E_G^{(2)}$ was derived (curve 3), and it is seen that it merges with the intrinsic curve in the metallic range.

This seems to add up to fairly strong evidence that there is a tail extending ~ 0.12 eV above the parabolic part of the valence band edge. There may also be a tail on the conduction band, but there is very little evidence bearing on this possibility. Such a tail may be contributing to the intermediate region between the extrinsic and intrinsic curves in Fig. 7.31.

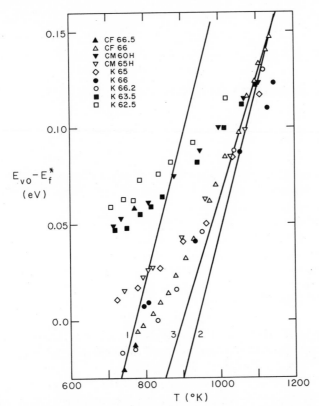

Fig. 7.31. Dependence of the apparent Fermi energy E_f^* on T for various compositions of $Tl_x Te_{1-x}$ for $x \gtrsim \frac{2}{3}$. Compositions are given in at% Tl, and the sources of the data are: CF (Field, 1967), CM (Cutler and Mallon, 1966), K (Kazandzhan *et al.*, 1972a). Curves 1 and 2 are plots of $-E_G^{(1)}$ and $-E_G^{(2)}$ versus T, respectively, and curve 3 is a plot of the calculated value of E_f^* in the metallic approximation with $E_G = E_G^{(2)}$.

The circled points in Fig. 7.22, provide information about ambipolar effects on the more sensitive parameter S in the composition range $0.48 < x < 0.60$. It is a simple matter to calculate $\xi_p = (E_{v0} - E_f)/kT$ from σ (Eq. 7.15), and then S_p (Eq. B4), and thus deduce $\Delta S = S_p - S$ for the points which deviate appreciably from the theoretical curve. The quantity $\ln(\Delta S \sigma)$ is plotted versus T^{-1} in Fig. 7.32 for two compositions, $x = 0.60$ and 0.50. In a two-band model, $\Delta S \sigma = \sigma_n (S_p - S_n)$ according to Eq. 6.25. With $E_{c0} - E_{v0} = E_G^{(2)}$, σ_n and S_n are calculated from Eqs. 7.4 and 7.5, yielding theoretical curves which are also plotted in Fig. 7.32. For $x = 0.60$, the theoretical curve for $\ln(\Delta S \sigma)$ is nearly parallel to the experimental curve, but $\Delta S \sigma$ is too large by a factor ~ 1.8. However $E_{c0} - E_{v0} < 0.12$ eV, and

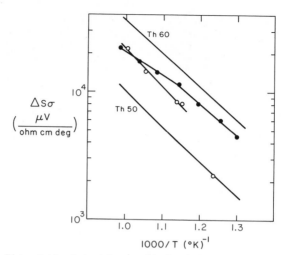

Fig. 7.32. Plots of $\Delta S\sigma$ derived from ambipolar deviations of S from the monopolar curve in Fig. 7.22. The theoretical curves (marked Th) are based on the Maxwell-Boltzmann approximation for S_n and σ_n, assuming that $E_G = E_G{}^{(2)}$. $\bigcirc$ = CM 60, $\bullet$ = 50. The code for the data sources is given in Fig. 7.31.

$E_{v0} - E_f \sim kT$. Therefore, the conduction band is overlapping the valence band tail, and the added transport due to the overlap (analogous to Eq. 7.10) would cause a contribution to $\sigma_n S_n$ which was not taken into account in the theoretical curve. This could account for the discrepancy.

In the case of $x = 0.50$, the experimental curve is ~ 2.8 times higher than the theoretical curve. In this case, our estimate $E_G{}^{(2)}$ suggests that $E_{c0} - E_f > 0.12$ eV over the entire range, and the only reasonable interpretation seems to be that the band gap is appreciably smaller than $E_G{}^{(2)}$. Assuming that $AC_n{}^2$ is unchanged, the magnitude of $\Delta S\sigma$ indicates that the band gap is ~ 0.076 eV smaller than $E_G{}^{(2)}$ on the average. If we ignore the relatively small change in $S_p - S_n$ with T, the slope of $\ln(\Delta S\sigma)$ indicates the activation energy for σ_n. Taking into account the dependence of E_f on T, we deduce an expression for the band gap $E_G{}^{(3)} = 0.44 - 5.4 \times 10^{-4}T$. It is not unreasonable for the band gap to change this much between $x = 0.6$ and $x = 0.5$, since the density of Tl–Te bonds changes from $0.6N_a$ to $0.5N_a$, according to our molecular model, and the density of Te–Te bonds changes from $0.1N_a$ to $0.25N_a$. The smaller value of E_G at $x = 0.5$ may reflect a change in the conduction band due to the replacement of antibonding states of Tl–Te bonds by those due to Te–Te bonds. Of course, one would expect the band structure near the conduction band edge to be strongly modified, so that $E_G{}^{(3)}$ can only be a rough estimate.

There has been some interest in the ambipolar contribution to the thermal conductivity, κ_m of Tl–Te. Mallon and Cutler (1965) found that $d\kappa/d(\sigma T)$ exceeds the metallic Wiedemann–Franz ratio $W_0 = (\pi^2/3)(k/e)^2$ by a large factor ($\gtrsim 1.5$) at compositions $x = 0.68, 0.54$, and 0.48. Fedorov and Machuev (1970a, b) have measured κ for $x = \frac{2}{3}, 0.5$, and 0.40, and their results indicate that $\kappa/\sigma T$ is significantly larger than W_0 only at $x = \frac{2}{3}$. For $x = 0.5$, κ_m can be written:

$$\kappa_m = \frac{\sigma_n\sigma_p}{\sigma} T(S_p - S_n)^2 \cong [\sigma_n(S_p - S_n)]\frac{(S_p - S_n)}{\sigma} \qquad (7.26)$$

since $\sigma_n \ll \sigma$, and the experimentally derived $\Delta S\sigma$ in Fig. 7.32 can be used to infer the quantity in the square bracket. The factor $S_p - S_n$ used in calculating the theoretical curve for $\sigma_n(S_p - S_n)$ in Fig. 7.32 is only 10–20% too large, judging by the difference between $E_G^{(2)}$ and $E_G^{(3)}$, so that its use in Eq. 7.26 leads to a negligible error in calculating κ_m. The values of ζ_p are also known, so that the accurate value of W, which is less than W_0, is obtained from the theoretical curve in Fig. 6.2 (marked κ_b) and used to determine the single-band electronic contribution $\kappa_p = W\sigma T$. The quantities κ_m and $\kappa_m + \kappa_p$ are plotted in Fig. 7.33. It is seen that $\kappa_m + \kappa_p$ agrees very well with the Fedorov–Machuev curve for $x = 0.50$, and shifting the theoretical curve upward, corresponding to an atomic contribution $\kappa_a = 1 \times 10^{-3}$ cal/cm deg sec yields perfect agreement. We show also the experimental curve of Mallon and Cutler for $x = 0.48$. The values of $\sigma(T)$ for this measurement agree very well with other measurements for $x = 0.50$, so that the correct composition is probably 0.50. The disagreement with the other curves in Fig. 7.33 indicates that there were systematic errors in these experimental results. The measurements by Mallon and Cutler at $x = 0.54$ and 0.68 are also too large, the latter exceeding the maximum $\kappa/\sigma T$ predicted by the overlapping band model in Fig. 6.2 by a large factor.

For the composition Tl_2Te, radiative contributions to κ have been suggested by two groups. As noted in Section 2.4, Dixon and Ertl (1971) have analyzed the effect of the sample length in their apparatus on the heat transport in Tl_2Te. They deduced values of the optical absorption coefficient $\alpha \sim 5$ cm^{-1}, leading to a radiation contribution $\kappa_r \sim \kappa/2$. Such small values of α are inconsistent with the magnitude of σ. In view of the discussion in Section 6.5, one expects the electrical conductivity at high frequencies $\sigma(\omega)$ to be at least as large as the DC conductivity. Since $\sigma > 70$ ohm^{-1} cm^{-1}, Eq. 6.38 leads to $\alpha \sim 10^4$ cm^{-1}, which indicates values of κ_r which are orders of magnitude smaller than κ (see Eq. 2.6). Fedorov and Machuev (1970b) have measured $\kappa(T)$ for Tl_2Te, and found that values of $\kappa/\sigma T$ are 2 to 3 times larger than W_0. They suggest that κ_r contributes to κ in addition

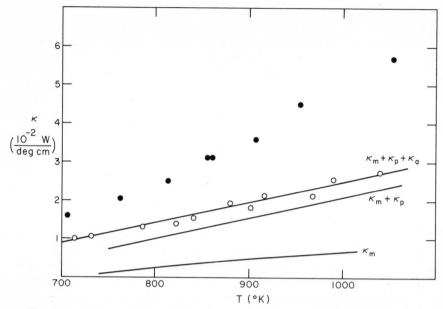

Fig. 7.33. Theoretical curves for κ_m, $\kappa_e (= \kappa_m + \kappa_p)$, and $\kappa_e + \kappa_a$ versus T for $Tl_{0.50}Te_{0.50}$ derived from values of $\Delta S\sigma$ shown in Fig. 7.32. The experimental data are from Fedorov (1970b) (○) and Mallon and Cutler (1965) (●). The assumed value of κ_a is 1.0×10^{-3} cal/cm deg sec.

to κ_m largely because they supposed the band gap to be too large to yield the required magnitude of κ_m. They apparently deduced the band gap from the activation energy of σ, but of course the activation energy is the band gap extrapolated to $T = 0$, and the band gap is actually very small at the experimental temperatures. Since the values of AC_n^2 and $A_pC_p^2$ are comparable in Tl_2Te, the theoretical symmetrical overlapping band model discussed in Section 6.3 can be compared with the experimental behavior of $\kappa_e/\sigma T$. In Fig. 7.34, we plot $(\kappa - \kappa_a)/\sigma T$ versus $-E_G/kT$ for the two cases $E_G^{(1)} = 0.58 - 7.5 \times 10^{-4}T$ and $E_G^{(2)} = 0.70 - 7.5 \times 10^{-4}T$, and compare them with the theoretical curves in Fig. 6.2 of the overlapping band model (marked κ_o). The value 4.2×10^{-3} W/deg cm was assumed for κ_a. It is seen that there is good agreement when the band gap has the larger value $E_G^{(2)}$. Figure 7.34 also shows the theoretical curve (κ_b^*) which ignores the extra contribution to $\sigma(E)$ due to the overlapping bands (Eq. 7.10), and it is seen that the curve which allows for the overlap contribution is in significantly better agreement with experiment. These results show that the observed behavior of κ is in very good accord with the electronic structure deduced for Tl_2Te, particularly with the wider band gap $E_G^{(2)}$, which was arrived at in this section.

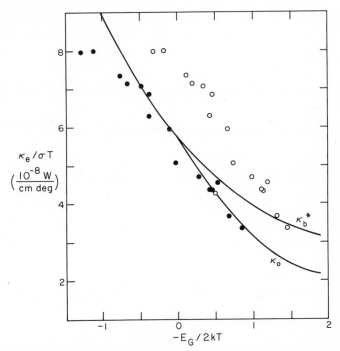

Fig. 7.34. Comparison of theory and experiment for $\kappa_e/\sigma T$ versus $-E_G/2kT$ using the symmetric parabolic band model. The points are calculated from $\kappa_e = \kappa - 1.0 \times 10^{-3}$ cal/cm deg sec from Fedorov (1970b), σ from Kazandzhan *et al.* (1972), and either $E_G^{(1)}$ (○) or $E_G^{(2)}$ (●) at various temperatures. The theoretical curve marked κ_b^* omits the contribution to $\sigma(E)$ which is proportional to $N_c N_v$, and the curve marked κ_0 includes this term.

7.5 MAGNETIC BEHAVIOR

The magnetic susceptibility χ of Tl–Te alloys has been measured at a number of compositions by Brown *et al.* (1971). They found a rapid increase in χ with T for $x < \frac{2}{3}$ which was ascribed to an increase in $N(E_f)$. In later work, more precise and detailed measurements have been made by Gardner and Cutler (1976), and these results were used to make a more specific analysis of the dependence of χ on $N(E_f)$, as inferred from the relation between χ and σ for $x < \frac{2}{3}$. In the metallic approximation, Eq. 6.31 is rewritten as:

$$\chi = \chi_d + b\mu_B^2 N(E_f) = \chi_d + b\mu_B^2 \sigma^{1/2}/A_p^{1/2}, \qquad (7.27)$$

where the constant b is introduced into the Pauli term to allow for the poorly understood many-body and band structure effects. The quantity b is expected to be of the order of one and nearly constant. The diamagnetic term χ_d includes the core and valence band contributions, and as discussed in Section 6.4.1, there are reasons to expect it to depend only on composition, although this is uncertain.

Figure 6.3 shows plots of χ versus $\sigma^{1/2}$ for several values of x in the Te–rich range of $Tl_x Te_{1-x}$. For $x > 0$, the points lie on a common straight line except for a small deviation at low σ for $x = 0.30$. This confirms Eq. 7.27, and the constant slope for $x \geq 0.10$ indicates that A_p is constant. The quantity χ_d seems not only to be constant with T, but apparently changes very little with x. For $\sigma \gtrsim 600$ ohm^{-1} cm^{-1}, the metallic approximation is poor, and expressions should be used which allow for the incomplete statistical degeneracy. This is done by replacing σ in Eq. 7.27 by σ^*, where:

$$\sigma^{*1/2} = -(A_p C_p^2)^{1/2} \int_0^\infty E^{1/2} (\partial f / \partial E) \, dE = (A_p C_p^2 kT/4)^{1/2} F_{-1/2}(\xi). \quad (7.28)$$

This equation is written in terms of hole energies. The function $\sigma^*(T)$ can be calculated from $\sigma(T)$, using Eq. 7.15 to determine $\xi = E_f/kT$. The quantity χ is plotted against $\sigma^{*1/2}$ in Fig. 7.35 for $x = 0.3$ and $x = 0.5$. Now a distinct bend is found in the curve for both compositions, with a constant slope at larger σ^* which corresponds to the one in Fig. 6.3. If this part of the curve is interpreted in terms of Eq. 7.27, the bend at lower σ^* indicates

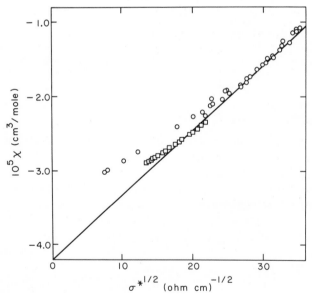

Fig. 7.35. Plot of χ versus $\sigma^{*1/2}$ from Gardner and Cutler (1976) for $Tl_x Te_{1-x}$ with $x = 0.30$ (circles) and 0.50 (squares). The solid line describes the behavior in accordance with Eq. 7.27.

an enhancement of the paramagnetic susceptibility at lower T, where E_f is small. At a given value of E_f this deviation is larger for $x = 0.3$ than for $x = 0.5$. It seems to indicate a very strong inverse dependence on T, since T is lower at the smaller x for the same E_f. Similar plots were made also at compositions $x = 0.60, 0.55, 0.45, 0.40$, and 0.35. None of them cover a wide enough range of σ^* to show a distinct change in slope that is seen for $x = 0.30$ and 0.50, but they fall into a pattern consistent with an enhanced paramagnetism which increases with decreasing E_f and T.

The reason for the enhanced paramagnetic susceptibility in Tl–Te is not yet established. Enhanced paramagnetism has been observed in heavily doped silicon, and some investigators believe that it is evidence of correlated electron gas behavior (Mott, 1972b; Berggren, 1974) in the impurity band. In the present case, the impurity band is absorbed into the valence band. It seems possible that the Hubbard interaction which causes the spin correlation may be strong enough to persist in Tl–Te even after the acceptor band is absorbed into the valence band, and manifests itself when E_f is close to the band edge. The states near the band edge are largely derived from the absorbed impurity band when E_f is near the band edge. However, further study will be necessary to evaluate this or other possible explanations.

Assuming that $b = 1$ in Eq. 7.27, the slope in Fig. 6.3 yields directly a value for A_p. It is convenient to express this in terms of the normalized transfer matrix element λ in Eqs. 6.14b and 6.11. The susceptibility curves yield $\lambda_p = 0.38$ This is in fair agreement with the result $\lambda_p \cong 0.22$ inferred in Section 7.3.3 from the IBM theory of bond equilibrium $(A_p = A_1 c^{*2}/c^2)$. A rougher value $\lambda_n = 1.1$ was derived from a comparison of χ and σ in the Tl–rich range (Gardner, 1976), which agrees well with the value $\lambda_n = 0.96$ derived from the rather accurate determination of A_n from transport data in Section 7.1. It is interesting to note that λ_n has a value ~ 1 in the conduction band, and it is much smaller (~ 0.3) in the valence band. This agrees with the expectation for bands whose atomic wave functions are s-like and p-like, respectively (Mott, 1969). For the same reason, the density of states parameter $C_p > C_n$. These two effects tend to cancel in $\sigma(E)$, so that $d\sigma/dE$ is nearly the same in the two bands. The large value of C_p indicates that the valence band is rather narrow (1–2 eV) and this is in accord with the observed deviation from a parabolic shape which starts at 0.25 eV below the valence band edge (Eqs. 7.14 and 6.14a).

The work of Brown et al. (1971) was mainly concerned with NMR measurements in $Tl_x Te_{1-x}$. They reported measurements of the tellurium Knight shift K_{Te} and spin–lattice relaxation times in the composition range $x < 0.69$. They found that $K_{Te} \propto \sigma^{1/2}$, in accord with Eq. 6.33, for a number of compositions. In view of the very good precision of the later measurements of χ, a more direct examination of the relation between K_{Te} and

$\chi = \chi_d + \chi_p$ implied by Eq. 6.33 can be made by plotting K_{Te} versus χ using the data of Brown *et al.* for K_{Te}, and the data of Gardner and Cutler for χ. The results are shown in Fig. 7.36 for $x < \frac{2}{3}$. The points tend to lie on line segments whose slopes indicate that P_f ($\propto dK/d\chi$) decreases at energies which lie deeper in the valence band. This suggests that the fractional s-character of the wave functions diminishes as the electron energy decreases. This is what one would expect if the s-character arises from admixture of the higher energy Te(6s) orbitals into the valence band, which, as discussed in Section 5.3, comes mainly from the Te(5p) states.

Measurements of K_{Tl} were made by Brown *et al.* (1971) and by Warren (1972b). They could be made only at compositions $x \gtrsim \frac{2}{3}$ because of the very large line width. There is not much data, but they show that K_{Tl} increases rapidly with x, for $x > \frac{2}{3}$, whereas K_{Te} is more nearly constant. Brown *et al.* note that there is an appreciable chemical shift which causes K_{Te} to be negative at $x \gtrsim \frac{2}{3}$. A chemical shift suggests the presence of relatively stable Tl$_2$Te molecules. Warren points out that the failure of K_{Te} to increase with x while K_{Tl} does increase is a further indication of the stability of Tl$_2$Te molecules. The difference in behavior indicates that the conduction band wave functions at E_f do not have an appreciable amplitude

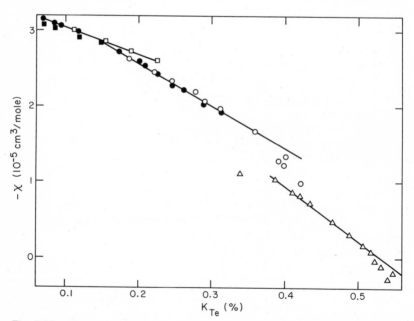

Fig. 7.36. Plot of the magnetic susceptibility (from Gardner and Cutler, 1976) versus the tellurium Knight shift (from Brown *et al.*, 1971 and Warren, 1972a) for various compositions of Tl$_x$Te$_{1-x}$ in the Te-rich range. At% Tl: 0($\triangle$), 20($\bigcirc$), 40($\bullet$), 55($\square$), 60($\blacksquare$).

at the Te sites. This is consistent with the genesis of the conduction band from Tl p orbitals and Tl–Te antibonding orbitals.

Measurements of the nuclear spin relaxation time have been reported for both Tl (Warren, 1972b) and Te (Brown, 1971). They indicate considerable enhancement η_R over the Korringa rate, but there does not seem to be quantitative agreement with Eq. 6.37.

7.6   PURE Te

The electronic and molecular structure of pure liquid Te is a keystone for understanding binary alloys containing Te in the limit of large concentrations of Te. Unfortunately liquid Te seems to be a very difficult material to understand, so that relatively little certain progress has been made in spite of the very many different types of measurements which have been made. Part of the difficulty is that its electrical properties stand at the borderline between those of typical metals and typical semiconductors (i.e., weak scattering and diffusive transport) as indicated by the fact that $2000 < \sigma < 3000$ ohm^{-1} cm^{-1} except for the relatively narrow range of T below 500°C.

A comprehensive examination of the basic character of molten Te has been made by Cabane and Friedel (1971). Their conclusions are based largely on the neutron diffraction studies of Tourand and Breuil (1970; Tourand *et al.*, 1972), discussed in Section 3.5, which indicate that Te is bonded largely in three-fold coordination at moderate temperatures. In terms of the bond orbital model discussed in Section 5.3, a third covalent bond causes an electron to occupy the antibonding σ^* states, which form the conduction band of Te. Consequently, Cabane and Friedel suggest that E_f is in the conduction band, which they note is in accord with the negative values of the Hall coefficient. They also make an extensive analysis of the Knight shift, and conclude that its magnitude is consistent with the value of $N(E_f)$ suggested by their model. It is suggested that the average coordination number z tends to decrease below 3 at $T < 500$°C (Cabane and Friedel, 1971; Tourand *et al.*, 1972), but this tendency is apparently less than originally supposed, since Tourand (1975) has more recently found that $z = 2.85$ at 403°C. Cabane and Friedel have ascribed the transition from two-fold to three-fold bonding with increasing T to an increase in entropy.

These conclusions are at considerable variance with the implications of the independent bond model for Tl–Te alloys. In the absence of Tl atoms, which act as chain stoppers, the IBM predicts chains of various lengths, terminated by dangling bonds. It ignores interactions between dangling bonds which certainly would occur when the chains are short, since the concentration of such species is normally very small. But for $x \cong 0$ and high T, this may not be the situation, particularly if energy for forming a dangling bond decreases at high values of c as suggested in Section 7.3.3. However, none of this takes into account the possibility of the three-fold bonding configurations which are indicated by the neutron-diffraction studies of Te.

The energy relations for three-fold bonding have been examined by Cutler (1976e). It was pointed out that, neglecting secondary bonding effects, a third bond is unstable if an electron must be promoted into the antibonding band, since the promotion energy cancels the bond energy, as discussed in Section 5.3. However, if dangling bonds are present as the result of broken Te–Te bonds, the extra electron of the third bond can go into an empty state in the valence band instead of the higher σ^* band, and three-fold bonds can then be stable. This suggests that the density of 3F (three-fold bonded) atoms will not exceed the density of 1F (dangling bond) atoms.

Consider next the energy required to form a 3F atom as compared to forming an ionized dangling bond. The latter was discussed in Section 7.3.2 in terms of two contributions: the energy E_1 required to create a neutral dangling bond and the energy E_- required to change it into an acceptor ion plus a hole. The quantity E_- was found in Section 7.3.2 to be relatively small, so that $E_d = E_1 \cong 0.25$ eV. A possible mechanism for forming a 3F atom is for a 1F atom to bind to a normally bonded 2F atom to form a 3F atom plus a 2F atom, with a net change 1F → 3F. This requires an energy change $E_3 = -2E_1 + E_+$, where E_+ is the electrostatic energy due to the extra positive charge on the 3F atom. In contrast to the small magnitude of E_-, E_+ can be expected to be relatively large at small hole concentrations. This is the result of nonlinear screening as illustrated in Fig. 7.37 in terms of the Thomas–Fermi model. Here, the TF radius r_0 is given by $4\pi r_0^3/3 = p^{-1}$. At low hole densities, the screening is nonlinear because the potential $V(r)$ is

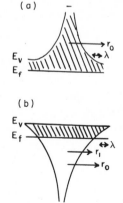

Fig. 7.37. Nonlinear screening of (a) a negatively charged acceptor ion and (b) a positively charged 3F atom according to the Thomas–Fermi model. The screening charge density is indicated by the shaded area between the valence band edge E_v and E_f. The quantity r_0 is the radius of the Wigner–Seitz sphere, λ is the linear screening distance, and r_1 is the radial distance from a 3F atom within which the screening charge density is constant (Cutler, 1976e).

comparable to $E_{v0} - E_f$. For a negative ion, the hole density becomes very large near the center. For a positive charge, the screening charge density is limited to the negative of the equilibrium hole density, so that the screening charge is uniform out to a distance $r \sim r_0$. This suggests a simple electrostatic model for calculating E_+ in which a positive charge is imbedded in a sphere of radius r_0 containing a uniform negative charge density. This yields:

$$E_+ \cong (e^2/K)(1/a - 1/r_0), \qquad (7.29)$$

where a is the radius of the central positive charge distribution and K is the dielectric constant. The appropriate value for K is nebulous, and $K = 5$ seems to be a conservatively large value. It seems reasonable to suppose that a is approximately the atomic radius ~ 2 Å, so that the first term in E_+ is greater than 1.30 eV. Therefore, when $r_0 \gtrsim 4$ Å (i.e., $c \gtrsim \frac{1}{8}$), E_+ is considerably larger than E_d, and $E_3 = E_+ - 2E_1 \gg kT$. At large values of c, E_+ decreases, and a transformation from 1F to 3F atoms can be expected when $E_3 \sim 0$. The transformation should occur relatively abruptly. We have suggested that it occurs when $\sigma \gtrsim 1500$ ohm^{-1} cm^{-1} (Cutler, 1976e).

Although a bond equilibrium theory for 3F bonding has not been worked out, we think that it would be difficult to account for the behavior of c versus x and T in $Tl_x Te_{1-x}$ for $x \gtrsim 0.2$ if the situation were that $E_3 < 0$, so that a large fraction of the dangling bonds would be transformed to 3F bonds. The change in electronic behavior between $x = 0.2$ and $x = 0$ suggests that a transformation to 3F coordination may be playing a role in this range. The results in Section 7.3.1 clearly show that E_f is in the valence band for $x \gtrsim 0.2$, and whatever else is happening, there is no evidence of a movement of E_f into the conduction band for $x < 0.2$ (see Fig. 7.21). This consideration seems to indicate that 3F bonding occurs by the 1F $\rightarrow$ 3F mechanism, rather than the one suggested by Cabane and Friedel. The function $N(E_f)$ may be as large as is indicated in their analysis, and the negative value of R_H would have to be explained by theoretical considerations such as those advanced by Friedman, as discussed in Section 6.2.

More recent diffraction studies (Tourand, 1975) indicate that $z \cong 3$ down to 400°C. This indicates that the large drop in σ (2500 to 1500 ohm^{-1} cm^{-1}) in this temperature range is not caused by a transition to two-fold bonding as supposed in earlier work. We think that a reasonable explanation for the behavior is possible if resonance (bond switching) is taken into account. According to this concept which was developed by Pauling (1960), there is a reduction in the energy for an electronic state which is a linear combination of states corresponding to a number of equivalent bond arrangements. One bond arrangement is shown in Fig. 7.38a for a liquid containing an equal number of 3F and 1F atoms, and it is apparent that many equivalent arrangements are possible by switching the bonds without moving the atoms. The

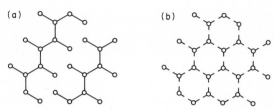

Fig. 7.38. Bond resonance in tellurium. (a) One of the equivalent bonding arrangements of 1F and 3F atoms. (b) The average bond configuration, with each pair of atoms connected by two-thirds of a bond.

overall electronic structure corresponds to two-thirds of a bond between each pair of atoms, as indicated in Fig. 7.38b. Since the role of 1F and 3F atoms is shared by the same atoms, there is no charge separation. Consequently the three-fold coordination is not inhibited by an electrostatic energy as described in Eq. 7.29. We suggest that such a resonating bond structure occurs in liquid Te at the lowest temperatures. The relatively low value of σ is caused by a small excess of 1F over 3F atoms, and an increase in this excess with T causes σ to increase. The resonance is strongly diminished when Te atoms are replaced in an alloy by nonequivalent atoms such as Tl. We suppose that this causes a shift toward two-fold coordination of Te atoms, and judging by the electrical behavior it occurs in the range $x \gtrsim 0.2$ in $Tl_x Te_{1-x}$.

We now consider the other implications of the electrical and magnetic measurements on tellurium. The curve for $\sigma(T)$ shows a rapidly rising part below 450°C, and a saturation at $\sigma \sim 2700$ ohm^{-1} cm^{-1} at $T \gtrsim 600$°C. This complete saturation is in contrast to the behavior of Tl–Te alloys as can be seen in Fig. 2.2b, and it suggests that $N(E_f)$ remains nearly constant. The behavior of S as σ changes (Fig. 7.21) is in accord with $\sigma(E_f) \propto E_f^{0.7}$ at $\sigma \gtrsim 2000$ ohm^{-1} cm^{-1}, with the exponent diminishing at larger σ, but evidence is lacking for the validity of a rigid band model. The function $\sigma(E)$ as well as E_f may be changing with T.

Further information comes from NMR and magnetic measurements (Cabane and Froidevaux, 1969; Warren, 1972a). Warren has analyzed the NMR behavior in relation to σ, and he finds that $\sigma \propto K^2$ except for a small deviation when σ exceeds 2500 ohm^{-1} cm^{-1}. This suggests that the diffusive transport mechanism applies up this value of σ. He has measured the magnetic relaxation rate, and found that $\eta_R \sigma \cong 2400$ ohm^{-1} cm^{-1}, but only for $\sigma \gtrsim 1500$ ohm^{-1} cm^{-1}. He also examined the dependence of K on χ, using the χ data of Urbain and Übelacker (1967) and noted a small decrease in $dK/d\chi$ at larger values of T. The plot of K versus χ in Fig. 7.36 which uses the χ measurements of Gardner and Cutler (1976) (which we think are more precise), shows a more pronounced decrease in slope, and it occurs at the temperature where $\sigma \gtrsim 2500$ ohm^{-1} cm^{-1}. Finally, a plot of χ versus $\sigma^{1/2}$, shown in Fig. 6.3, shows also a deviation from a straight line at $\sigma \gtrsim 2500$ ohm^{-1} cm^{-1}. It also has a smaller slope at $\sigma < 2500$ ohm^{-1} cm^{-1} than in Tl$_x$Te$_{1-x}$. This indicates that the parameters A_p in Eq. 7.27 and λ in Eq. 6.14b are larger, as one would expect with a higher coordination number.

Of course, the diffusive mechanism no longer applies when $\sigma > 2500$ ohm^{-1} cm^{-1}, and this can account for changes in $dK/d\sigma^{1/2}$ and $d\chi/d\sigma^{1/2}$. Thermal scattering may become significant, so that σ would decrease with T, but the change in $dK/d\chi$, which indicates a decreasing value of P_f, suggests that something more profound is also occurring. Perhaps the thermal expansion of the liquid causes a change in the band structure. Because of the small value of $d\sigma/dT$ in this range, a small $\Delta\sigma$ corresponds to a large ΔT. A change in the structure of the liquid in the direction of six-fold coordination (like crystalline polonium) has been suggested by Tourand *et al.* (1972). No explanation is evident for the decrease in $\sigma\eta_R$ for $\sigma \gtrsim 1500$ ohm^{-1} cm^{-1}. This seems to be a manifestation of a more general deviation from Eq. 6.36 already noted, which is apparent in the results of Brown *et al.* (1971) for Tl$_x$Te$_{1-x}$.

CHAPTER 8

OTHER ALLOYS

In this final chapter we discuss a number of liquid semiconductor systems other than Tl–Te. Much of what follows will be tentative or speculative because of the poorly developed status of the subject. Before discussing specific systems, it will be useful to consider the behavior of binary alloys from a general point of view.

8.1 GENERAL BEHAVIOR

8.1.1 Bonding

In Chapters 2 and 3 it was observed that properties of many binary alloys M_xA_{1-x} indicate a characteristic composition M_aA_m corresponding to a compound in which M and A exhibit common chemical valences m and a, respectively. (We adopt the convention of letting M refer to the more electropositive constituent.) In this section we confine our attention to such alloys, and discuss in a general way the molecular and the electronic structure at the stoichiometric composition, and the effects of deviations from stoichiometry in the directions of excess M or A.

Whenever available, the thermochemical information discussed in Section 3.4 indicates that a chemical compound forms at the stoichiometric composition. However, the binding may range between ionic and covalent, and there seems to be considerable diversity of opinion about the ionicity of various compounds. There is little direct experimental information on this question. Enderby and coworkers (Hawker *et al.*, 1973) have suggested an ionic structure for Cu_2Te on the basis of their interpretation of neutron diffraction results. Warren (1973a) has derived the same conclusions about CuTe from a study of NMR data, but he has found evidence of somewhat less ionicity in Cu_2Te. In the case of $Ga_2(Se_xTe_{1-x})_3$, Warren (1973) has found evidence that the bonding is covalent.

There is a tendency to regard ionic bonding as more consistent with fluidity than covalent bonding, particularly when elements are present with a valence of three or more. However an infinite bonding network is consistent with fluidity if the bond lifetimes are short, and this situation is indicated in some alloys by results of NMR studies. Warren (1972b) has deduced from quadrupole relaxation rates that the bond lifetime $\tau_b \sim 10^{-11}$ sec in Ga_2Te_3. Since the vibration period $\tau_v \sim 10^{-13}$ sec, the relation $f \sim \tau_v/\tau_b$ indicates that a fraction $f \sim 1\%$ of the bonds are dissociated at a given time. This is consistent with considerable fluidity for the liquid. Much longer periods for atomic migration $\sim 10^{-6}$ sec have been found in molten Se from the analysis of NMR line shapes (Bishop, 1973), but this is also consistent with the larger viscosity which is observed. In extreme cases where f becomes very small as in As_2Se_3 and related liquids, very large viscosities occur which leads to the formation of glasses.

A general guide to the ionicity of bonds is provided empirically through the difference in electronegativity $\Delta X = X_A - X_M$ (Pauling, 1960). Electronegativity is a measure of the strength of attraction of an element for a valence electron, and it has been defined in various ways. There is a large literature on the subject which reflects considerable disagreement on the best way to determine the electronegativity. It is inexactly defined, but nevertheless it is a real property of an element, and tables compiled by various workers seem to lead to similar predictions for the values of ΔX. We list in Table 8.1 electronegativity values taken from the work of Gordy and Thomas (1956). Table 8.2 contains values of ΔX for various compounds

TABLE 8.1

Electronegativity Values from
Gordy and Thomas (1956)

Element	X	Element	X	Element	X
Li	0.95	Zn	1.5	I	2.6
O	3.5	Ga	1.5	Cs	0.8
F	3.9	Ge	1.8	Ba	0.9
Na	0.9	As	2.0	Au	2.3
Mg	1.2	Sr	1.0	Hg(II)	1.8
Al	1.8	Ag	1.8	Tl(I)	1.5
P	2.1	Cd	1.5	Tl(III)	1.9
S	2.5	In	1.5	Pb(II)	1.6
Cl	3.0	Sn(II)	1.7	Se	2.4
K	0.8	Sn(IV)	1.8	Br	2.8
Ca	1.0	Sb(III)	1.8	Rb	0.8
Cu(I)	1.8	Sb(V)	2.1	Pb(IV)	1.8
Cu(II)	2.0	Te	2.1	Bi	1.8

TABLE 8.2

Bond Electronegativity Differences ΔX and
Ionicities I of Liquid Semiconductor Compounds

Compound	ΔX	I	Compound	ΔX	I
Cu_2Te	0.3	0.022	Mg_3Bi_2	0.6	0.086
$CuTe$	0.1	0.0025	$PbSe$	0.8	0.148
Ag_2Te	0.3	0.022	$SnSe$	0.7	0.115
$SnTe$	0.4	0.039	In_2S_3	1.0	0.22
$PbTe$	0.5	0.061	Tl_2Te	0.6	0.086
$ZnTe$	0.6	0.086	Tl_2Se	0.9	0.183
$CdTe$	0.6	0.086	Tl_2S	1.0	0.22
Ga_2Te_3	0.6	0.086	$CsAu$	1.55	0.45
Ga_2Se_3	0.9	0.183	Li_3Bi	0.85	0.165
In_2Te_3	0.6	0.086	As_2Te_3	0.1	0.003
Sb_2Te_3	0.3	0.022	As_2Se_3	0.4	0.039
Bi_2Se_3	0.6	0.086	$NaCl$	2.2	0.70
Bi_2Te_3	0.3	0.022			

of interest as liquid semiconductors. We are interested in the bond ionicity I, which can be defined (Pauling, 1960) as p/ed, where p is the dipole moment of the bond and d is the bond distance. Pauling has given an empirical formula relating I and ΔX:

$$I = 1 - \exp(-\Delta X^2/4), \qquad (8.1)$$

and values of I determined in this way are also listed in Table 8.2. It is to be noted that in most cases the ionicities are rather small. For reference, $I = 0.70$ for NaCl. In particular, $I \sim 0.02$ for Cu_2Te, which compares with 0.09 for Tl_2Te and Ga_2Te_3, suggesting that Cu_2Te is bonded covalently. However, the possibility of formation of molecular complexes complicates the evaluation of the ionicity. If, for instance, Cu_2Te molecules maintain a resonating configuration where the bonds switch back and forth, such as is shown in Fig. 8.1, the ionicity is greatly increased. The complexing ratio R is defined as the ratio of the coordination number of an atom to its valence, and the effective ionicity I_c is given by (Phillips, 1973):

$$I_c = 1 - (1 - I)/R. \qquad (8.2)$$

It is seen that a two-fold complexing ratio for Cu_2Te would raise the effective ionicity to a value greater than 0.5. It is hard to determine R for a

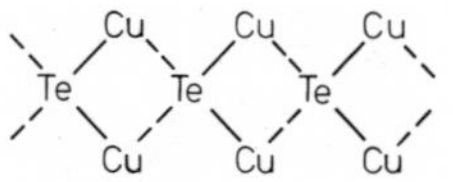

Fig. 8.1.   Complexing in Cu_2Te.

liquid, in contrast to a crystalline solid. Obviously thermal disorder tends to keep it from having a large value.

Let us now consider the molecular structure for nonstoichiometric compositions. In the case of Tl–Te, it was found that when the excess element is chemically bond^d (Tl_2Te + Te), the expected electronic behavior is profoundly different than when it is not (Tl_2Te + Tl). A strongly ionic M–A bond does not preclude covalent bonding of excess A. It seems possible that Na–Te alloys, for instance, contain ionic chain molecules of the form $(Te_n)^{2-}$ in the Te-rich range. The best guide seems to be whether or not the excess element bonds covalently with itself. This suggests that elements in groups IVB, VB, and VIB of the periodic table, particularly the lighter ones, may bond covalently when in excess. Heavy elements in groups IV and V such as Sn or Bi seem to act as pseudogroup II or III elements and bond metallically when in excess. On the other hand, the electronic behavior of As–Se alloys, discussed in Section 8.8, suggests that both As and Se are covalently bonded when in excess of the composition As_2Se_3 (Hurst and Davis, 1974a, b). Again, the electrical behavior of many Te and Se alloys suggests covalent bonding of the excess chalcogenide.

Another type of situation occurs when an element has more than one valence, since an excess of that element may bond with a mixed valence. For example, In has valences 1 and 3, with the latter preferred (judging by the relative number of common compounds), and Cu has valences 1 and 2 with nearly equal preference. Tl also has a valence of 3 in addition to the preferred value of 1. In some cases, the behavior of binary alloys containing these elements may reflect bonding with mixed valency. This possibility is best discussed in later sections where specific cases are considered (see Section 8.4).

8.1.2 Electronic Structure

There seems to be general agreement that the electronic structure at the stoichiometric composition is characterized by a pseudogap. We have found it convenient to describe the pseudogap in terms of two bands, with a band gap which may be small or negative. When it comes to describing the electronic structure of alloys at nonstoichiometric compositions, however, there has been a considerable diversity of views. Aside from models based on a heterogeneous microstructure, which we think are contradicted by thermodynamic considerations (see Section 3.4), a number of models have been advanced for the electronic structure of liquid semiconductors which are reviewed here. The models of other authors have all taken the view that for both M and A the excess element is in an unbonded, atomic form.

Enderby and Collings (1970) have suggested a model for M_xA_{1-x} in which an excess of either M or A over the stoichiometric composition causes the valence electrons of the excess atoms to go into a rigid conduction band, as indicated in Fig. 8.2. On changing composition from excess M to excess A, E_f decreases to a minimum (determined by the equilibrium concentration of the dissociated compound) and then increases. In order to explain the positive thermopower which is frequently found with excess A, they postulate a relatively narrow resonant scattering impurity state in the band just below E_f. This model has been criticized for not explaining the temperature dependence of S, and for the fact that very special behavior of the resonance scattering is required to cause S to change sign at the stoichiometric composition (Faber, 1972). We also think that it is unnatural for E_f to increase with respect to the existing bands as the result of adding a constituent with more than the average electronegativity.

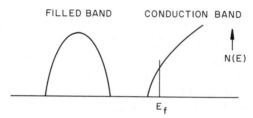

Fig. 8.2.   Enderby's model for the electronic structure of an M–A alloy. E_f always remains in the conduction band, and bonding electrons go into the filled band.

Faber (1972) has modified this model by assuming that the excess element forms an impurity band below the bottom edge of the conduction band, as shown in Fig. 8.3. This avoids the difficulty with respect to the thermopower. With excess M, E_f will be below the maximum of the impurity band, and with excess A, E_f will be above the minimum in typical alloys. Transport is in an impurity band near the stoichiometric composition, and the impurity band grows into a full-fledged band with large concentrations of excess M or A. This model would explain the observed behavior qualitatively, but we question whether it can provide a quantitative explanation in many cases. For reasons discussed in Section 7.2, an appreciable band gap is necessary if there is to be an impurity band which is not absorbed into the conduction band. We think that a distinct impurity band is unlikely unless the minimum conductivity is quite a bit less than 100 ohm^{-1} cm^{-1}. If transport is in an impurity band at compositions near stoichiometry, the density of states should be very sensitive to the composition, and a rigid band model would be bad. But in the case of Tl–Te, we have found that rigid band behavior occurs on both sides of the stoichiometric composition.

Another type of model for the electronic structure, proposed by Roth (1975), is illustrated in Fig. 8.4. When small amounts of A are added to

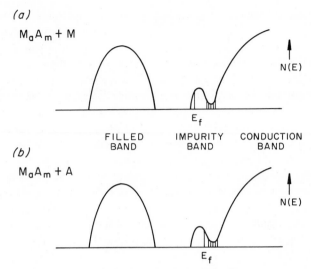

Fig. 8.3.   Faber's model for the electronic structure of an M–A alloy. E_f is always in the impurity band, which grows with increasing concentration of excess A or M. The hatch marks indicate a possible mobility gap between the impurity band and conduction band.

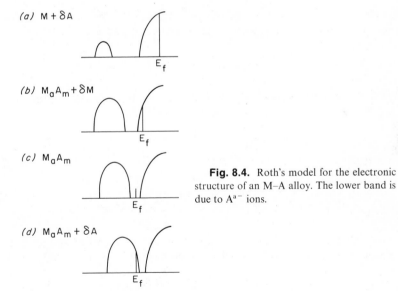

Fig. 8.4.   Roth's model for the electronic structure of an M–A alloy. The lower band is due to A^{a-} ions.

M, bonding electrons go into states which are created below E_f and may or may not be separated from the conduction band by a gap. This new band grows with decreasing x, and E_f moves downward in the conduction band since each A atom adds more states than electrons to the lower band. At the stoichiometric composition, E_f is near the center of the pseudogap. An excess of A causes E_f to drop into the valence band, and as the composition changes towards A, the valence band becomes a partially filled band of the pure A constituent. This model assumes completely ionic bonding, although a slight modification discussed below would make it applicable to covalent bonding between M and A.

We think that Roth's model is appropriate in the limit of ionic bonding. Ionic and covalent bonding can be regarded as extreme limits of a continuous change in the character of the bond which occurs as a function of the difference in electronegativity ΔX. To understand the relation between these extremes, let us reexamine the bond orbital model for the general case where there is a difference in electronegativity.

In Section 5.3 we have discussed a simple molecular orbital model for covalent bonding. It was assumed there that the atomic orbitals have the same initial energy. If the atoms are different, the initial energies of the atomic orbitals will be different, the electropositive element having the higher energy. In this case the bond will be partially ionic. Following Coulson (1961), the wave function formed from orbitals ψ_M and ψ_A is of the form:

$$\psi = \frac{\lambda}{(1 + \lambda^2)^{1/2}} \psi_A + \frac{1}{(1 + \lambda^2)^{1/2}} \psi_M, \tag{8.3}$$

with $\lambda > 1$ for the bonding (σ) orbital and $\lambda < 1$ for the antibonding (σ^*) orbital. In the bond orbital scheme, the ionicity is given by:

$$I = (1 - \lambda^2)/(1 + \lambda^2). \tag{8.4}$$

The splitting $2D$ between the σ^* and σ energies is related to the energy difference $2\Delta = (E_M - E_A)$ between the energies of the atomic orbitals and the interaction integral $H = \langle \psi_M | \Delta V | \psi_A \rangle$ by:

$$D = (\Delta^2 + H^2)^{1/2}. \tag{8.5}$$

It is clear that $D \to \Delta$ in the ionic limit where $\Delta \gg H$. Corresponding to this effect on the splitting, the ionicity is increased through the relation:

$$\lambda_\sigma^{-1} = \lambda_{\sigma^*} = (D - \Delta)/H. \tag{8.6}$$

It should be noted, however, that the charge transfer due to increasing ionicity shifts the potentials of the two atoms toward each other, and the final solution must be a self-consistent one in the same spirit as the Hartree solution for the electronic structure of an atom. Rough calculations indicate that appreciable shifts in energy can occur which would cause Δ to be very small, in response to relatively small deviations of λ from 1 (i.e., small ionicities). But if the original difference in energies of the atomic orbitals of the uncharged atoms is large enough, the final bond is strongly ionic. We show this schematically in Fig. 8.5, where the shifts in E_M and E_A due to charging are indicated as well as the splitting due to bonding, for the case of large ionicity. If the A atom is tellurium, comparison with the homopolar case in Fig. 5.4b shows that the main difference is that the nonbonding π electrons are close in energy to the σ bonding electrons. Assuming tight-binding bands, the π and σ bands would overlap strongly in energy and would presumably be merged. Conversely, the antibonding band is close in energy to the original M atomic orbitals.

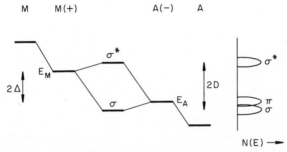

Fig. 8.5. Energy level scheme for partially ionic bonds. The M and A levels are shifted to M(+) and A(−) as a result of the polarity of the M–A bonds. The tight-binding bands are shown on the right, assuming that the A atoms have some nonbonding states.

To see how covalency modifies the electronic structure in comparison with the ionic limit of Roth's model, we show in Fig. 8.6 the same sequence for the case where the M–A bond is partially covalent, and A is a chalcogenide element. For a small concentration of A in M (Fig. 8.5a) the lower A band is split between the σ states and the somewhat higher π states. The antibonding σ^* states must be above E_f in the conduction band if there is to be bonding. In accordance with Eqs. 8.3 and 8.4, partial covalent bonding causes removal of M states from the conduction band below E_f to the extent that I is not large. Near the stoichiometric composition (Fig. 8.6b), the σ, π, and σ^* bands have grown and the original M band has diminished, and as discussed in Section 7.2, the M band becomes an impurity band which is either in the band gap between π and σ^*, or is absorbed into the σ^* band. The electronic structure at M_aA_m is shown in Fig. 8.6c. It is seen that the partial covalency causes no qualitative changes from ionic bonding up to this point.

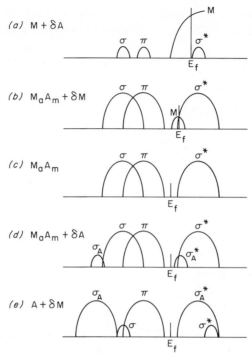

Fig. 8.6.   Model for the electronic structure of an M–A alloy, assuming that excess A bonds covalently to itself. (a), (b), and (c) differ from the ionic model of Fig. 8.4 only in a partial separation between the σ bonding and π nonbonding states of the valence band. In (d) and (e) the excess A gives rise to σ_A bonding and $\sigma_A{}^*$ antibonding states instead of adding electrons and states to the π band.

On going to excess A, however, there is a qualitative difference if the A molecules bond covalently to each other. Assuming that the M atoms have only one valence, the excess A atoms must either bond to other A atoms, which means that the bonding is purely covalent, or else it takes an ionic form, probably sharing its charge with the other A atoms. In the latter case, the sequence is the same as in Roth's model (Fig. 8.4d and e), where E_f moves downward in the valence band. If the A atoms bond to each other, as indicated in Fig. 8.6d and e, the valence band remains filled and insulator behavior occurs unless some covalent bonds are broken.

Figure 8.6 describes what we think occurs in most binary alloys in which A is Te or Se. If the electronegative constituent is a group VB element, there is no π band. The electrical behavior of Mg–Bi seems to indicate that the excess Bi does not bond covalently and the ionic description (Fig. 8.4) seems to be appropriate. We shall refer to these models (Fig. 8.4 or 8.6) as the superposed band model.

8.2   Mg–Bi AND SIMILAR ALLOYS

Liquid semiconductor behavior has been observed in several binary alloys whose electronegative constituent is more electropositive than tellurium: Mg_3Bi_2, Li_3Bi, and $CsAu$. In each case, both elements are conventional liquid metals. As we shall see, their electrical properties have other similarities which suggest that they should be considered together.

Mg_xBi_{1-x} was first studied by Illschner and Wagner (1958), who observed a sharp minimum in $\sigma(x)$ at the composition Mg_3Bi_2. Enderby and Collings (1970) confirmed the behavior of σ, and also measured $S(x)$. These curves are shown in Fig. 8.7. The shape of $S(x)$ is similar to Tl_xTe_{1-x}, but it differs quantitatively in having much smaller extrema (10 μV/deg) which occur at a much larger displacement from the stoichiometric composition ($\Delta x \sim 0.1$ instead of ~ 0.01).

The behavior of $S(x)$ could be explained in terms of a strong overlap between the valence and conduction bands. However, the band gap must be positive and of the order of kT (~ 0.1 eV) at $x = 0.6$ since the minimum σ (45 ohm^{-1} cm^{-1}) is smaller than the estimated value at a mobility edge (~ 200 ohm^{-1} cm^{-1}). Therefore the width of the pseudogap must change with x, since $S(x)$ would be quantitatively similar to Tl–Te with extrema of $\gtrsim 100$ μV/deg if the band gap remained positive.

A reasonable explanation of the behavior seems possible in terms of the superposed band model if it is assumed that there is an appreciable degree of ionicity. With partially ionic bonding in a compound M_aA_m, each bond dipole places an effective charge $z_M = aI$ on the M atom and $z_A = mI$ on the A atom. This causes a potential difference which can be conveniently thought of as a Madelung potential. There is no lattice, of course, but if the potential is expressed in terms of the interatomic distance, then the Madelung constant for binary salts is nearly independent of the lattice (Tosi, 1964). We can conveniently use this result, which leads to a Madelung potential:

$$\phi = 0.8(a + m)e^2 I^2 / r \cong 2.9(a + m)I^2 \quad \text{eV}, \qquad (8.7)$$

where r is the interatomic distance. The second expression assumes $r = 4$ Å, which is close to the true value for most liquid semiconductors. Table 8.2 gives $I = 0.086$ for Mg_3Bi_2, which implies $\phi \sim 0.1$ eV, but if complexing occurs in a fourfold coordination, $I_c \sim 0.3$ and $\phi \sim 1$ eV.

When partial ionic bonding occurs, the conduction band wave functions are concentrated on the M atoms, and the valence band wave functions are concentrated on the A atoms if the top of the valence band is derived from bonding electrons (see Eqs. 8.3 and 8.4). Therefore, the effect of the Madelung potential is to increase the band gap by raising the conduction band with respect to the valence band. If a high concentration of electrons or holes is created as the result of excess M or A which is not covalently bonded, the increased screening will sharply reduce ϕ, and cause the band gap to decrease.

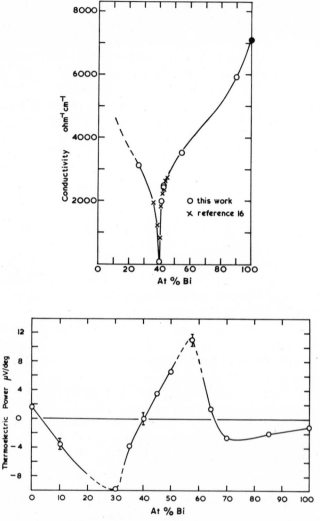

Fig. 8.7. Isotherms for the electrical conductivity and the thermopower for Mg–Bi alloys. $\sigma(x)$ is at 900°C (except for one composition) and $S(x)$ is at 100°C above the liquidus temperature (Enderby and Collings, 1970).

We think that this happens in $Mg_x Bi_{1-x}$, and we assume that $\phi > 0.1$ eV at the stoichiometric composition. The shape of $\sigma(x)$ seems consistent with the hypothesis, since σ drops sharply within a very narrow range of composition about $Mg_3 Bi_2$, as would occur if the band gap increased abruptly in this range.

Liquid semiconductor behavior was predicted in Li_xBi_{1-x} near the composition Li_3Bi (Ioannides *et al.*, 1973) and measurements at $x > 0.73$ have confirmed this (Steinleitner *et al.*, 1975b). The quantity σ drops to a minimum (~ 500 ohm^{-1} cm^{-1}) in a relatively narrow range of compositions near $x = 0.25$, and $d\sigma/dT$ is positive in the same range ($\Delta x \sim 0.05$). The peak value of $d\sigma/dT$ at $x = 0.2$ corresponds to an activation energy ~ 0.18 eV. The excess volume of mixing ΔV was also measured, and the peak of -37% at $x = 0.25$ is noteworthy for its unusually large magnitude and its negative sign. Other liquid semiconductors such as In_2Te_3 and Tl_2Te have positive values of $\Delta V/V \sim 10\%$. This different behavior for Li_3Bi may reflect bonding which is ionic rather than covalent, and the peculiar nature of the Li atom (i.e., its small 1s core).

The compound CsAu may seem like an unlikely candidate for a liquid semiconductor, but it has the largest ionicity of all the compounds listed in Table 8.2. We show $\sigma(x)$ for Cs_xAu_{1-x} in the reported range $x > 0.43$ (Hoshino *et al.*, 1975 and Schmutzler *et al.*, 1976a, b) in Fig. 8.8, and it is seen that the minimum value at $x = 0.5$ is $\sigma \sim 3$ ohm^{-1} cm^{-1}. As in the case of Li_xBi_{1-x},

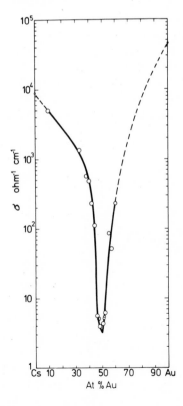

Fig. 8.8. The conductivity σ of Cs–Au alloys at 600°C as a function of the concentration (Schmutzler *et al.*, 1976b).

$d\sigma/dT$ is positive only within a few atomic percent of the stoichiometric composition. The peak magnitude corresponds to an activation energy ~ 0.23 eV. The authors note that $d\sigma/dT$ becomes positive at the composition where σ drops below 100 ohm^{-1} cm^{-1}. Aside from the difference in the magnitude of σ, the behavior of Cs_xAu_{1-x} is very much like Li_xBi_{1-x}, and we think that both reflect the formation of an ionic compound. The ionic model in Fig. 8.4 seems appropriate for both systems. In the case of CsAu, it seems likely that the ionic compound Cs^+Au^- is formed. The energy of the valence 6s shell of gold is low compared to the other noble metals (due to the strong spin–orbit interaction). It seems likely that this difference, together with the Madelung potential, causes the filled Au $(6s^2)$ band of the compound to be lower in energy than the conduction band which may arise from the Cs 6s and the Au 6p states.

The very low minimum σ (~ 3 ohm^{-1} cm^{-1}) seems too small to be consistent with the activation energy E_A (~ 0.23 eV) for normal transport at the mobility edge. Since the band gap E_G normally has a negative temperature coefficient, we would expect that $E_G < 2E_A$. On the other hand, excitation across the band gap gives $\sigma > 2\sigma(E_c) \exp(-E_A/kT)$, and $\sigma(E_c) \sim 200$ ohm^{-1} cm^{-1} leads to 25 ohm^{-1} cm^{-1} as a lower limit for σ. It therefore seems likely that the strong ionic character of the compound causes excess holes or electrons to be immobilized in the form of small polarons.

The hypothesis that electrons or holes are trapped at compositions near CsAu receives support from the behavior of the thermopower, shown in Fig. 8.9 (Schmutzler et $al.$, 1976a, b). Instead of changing sign near the stoichiometric composition with characteristic peaks in the magnitude at

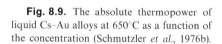
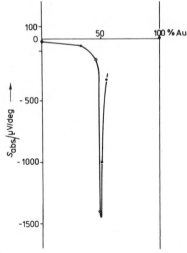

Fig. 8.9. The absolute thermopower of liquid Cs–Au alloys at 650°C as a function of the concentration (Schmutzler et $al.$, 1976b).

neighboring compositions, S remains negative with a very deep minimum at $x = 0.5$. The original workers note that the very great magnitude ($\gtrsim 1000 \ \mu V/deg$) is inconsistent with electronic transport and must be ascribed to ionic transport.

A study has also been made of the magnetic susceptibility of Cs_xAu_{1-x} (Steinleitner and Freyland, 1975). As one would expect, there is a large diamagnetic deviation from the average for the two metals which peaks at $x = 0.5$. An effort was made to tie the magnitude of the deviation to the presence of Au^- ions, but definite conclusions were not possible.

8.3   Ag–Te ALLOYS

Silver is chemically very similar to thallium. They are both monovalent and they tend to form chemical compounds with similar properties. Therefore, a comparison of the behavior of Ag–Te alloys with that of Tl–Te provides a good test for the model proposed in Chapter 7 for the behavior of Tl–Te.

The main source of experimental information is in the work of Dancy and Derge (1963), and there are also some unpublished data from the laboratory of the author (Cutler, 1963). We show Dancy's isotherms for σ and S in Fig. 8.10, and they can be seen to be very similar in shape and magnitude to the isotherms for Tl–Te (Fig. 2.3). The data can be used to make more detailed comparisons with Tl–Te in the Te–rich range. This is of particular interest since the independent bond model proposed for Tl–Te should also apply to Ag–Te. Since the Ag or Tl plays the passive role of a chain stopper, the concentration of dangling bonds in the IBM should be the same function of temperature and composition for both alloys. The function $\sigma(x)$ for the two alloys at $T = 800°K$ are compared in Fig. 8.11. They agree within the probable experimental error for $x > 0.4$, but the Ag points rise to higher values at $x < 0.3$. In Fig. 8.12, $\ln \sigma$ is plotted versus T^{-1} for Ag_xTe_{1-x} for several compositions, and comparison is made with a typical Tl curve. The shapes are similar, and the slopes are only slightly larger for the Ag curves. Since the differences between Ag and Tl are small enough to be attributed to secondary factors, these observations support the validity of the molecular bond model.

The similar magnitudes of S and σ for the two alloys indicates that the valence band structure may be similar. Therefore we have plotted $\ln \sigma$ versus $\ln(S/T)$ for Ag_xTe_{1-x} in Fig. 8.13 and make a comparison with the curve for Tl–Te (Fig. 7.22). The deviations of points for $x > 0.4$ from a common curve seem to be larger than the probable experimental error, and they are in the wrong direction to be caused by ambipolar transport. Therefore, it seems likely that the rigid band model fails in this composition range. The

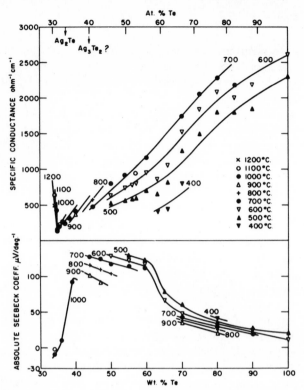

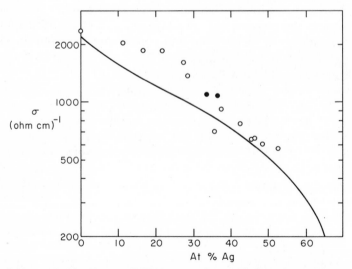

Fig. 8.10. The electrical conductivity and thermopower of the molten Ag–Te system (Dancy, 1965).

Fig. 8.11. $\sigma(x)$ of Ag_xTe_{1-x} at 800°K in comparison to Tl–Te. Sources of data are Dancy (1965) (○) and Cutler and Leavy (●) (Cutler, 1963).

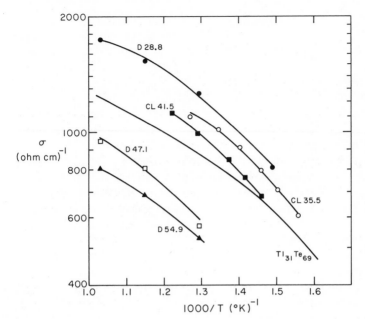

Fig. 8.12. Comparison of ln σ versus T of Ag_xTe_{1-x} at various compositions (given in at% Ag) in comparison with a typical curve for Tl–Te. Sources of the data are D (Dancy, 1965) and CL (Cutler, 1963).

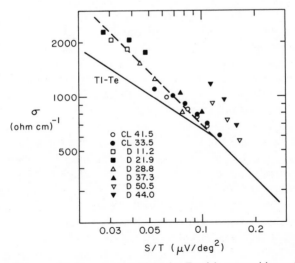

Fig. 8.13. The quantity ln σ versus $\ln(S/T)$ for Te–rich compositions of Ag_xTe_{1-x} in comparison with the curve for Tl–Te in Fig. 7.22. Compositions are given in at% Ag. Sources of data are D (Dancy, 1965) and CL (Cutler, 1963).

points at lower x lie near a common curve which seems to be identical with the straight line (with a slope -1) which describes the behavior of Tl–Te for $\sigma \gtrsim 700$ ohm^{-1} cm^{-1}. In the present case, it would seem that $\sigma(E) \cong 2960E$ ohm^{-1} cm^{-1} eV^{-1} is valid up to $\sigma \sim 2000$ ohm^{-1} cm^{-1}. The difference in $\sigma(x)$ for silver and thallium which is seen in Fig. 8.11 at $x \gtrsim 0.4$ can be traced to the negative deviations from this relation for Tl.

The electronic structures of Ag and Tl atoms differ in that the filled 4d shell of Ag is close in energy to the 5s state, in comparison to the corresponding states (5d and 6p) for Tl (Herman and Skillman, 1963). As a result, the high-lying 5d states of the Ag atoms probably overlap the valence band, and this may account for the differences noted in the electronic structure. At high x, the varying contribution of the d states to the valence band may cause $N(E)$ near the band edge to change, causing nonrigid band behavior in the Ag alloy. Even at lower Ag concentrations, the 4d contributions may possibly help to preserve the parabolic behavior of $N(E)$ at larger distances from the band edge than is observed in Tl–Te.

These conclusions obviously must be very tentative, and more extensive and precise data would clearly help to make a comparison with Tl–Te more definitive.

8.4 Cu–Te ALLOYS

This system is of special interest because of its similarities and differences with respect to Ag–Te and Tl–Te. The main source of information on the transport properties is in the work of Dancy (1965a), and her isotherms for σ and S are shown in Fig. 8.14. There are also some unpublished thermoelectric measurements (Cutler, 1963). Warren (1973a) has made an NMR study and his isotherms for the Knight shift are shown in Fig. 8.15. In addition, the molecular structure of CuTe and Cu$_2$Te has been studied by neutron diffraction (Enderby and Hawker, 1972; Hawker *et al.*, 1973).

Superficially, the behavior of S and σ in Fig. 8.14 resembles Tl–Te (Fig. 2.3) and Ag–Te (Fig. 8.12), but careful examination reveals very distinctive differences. At $y < 0.33$ in Cu$_{1-y}$Te$_y$, σ increases slowly with excess Cu and rapidly with temperature, which is opposite to the behavior in Ag and Tl alloys. Furthermore, the thermopower is positive and has a maximum at $y = \frac{1}{3}$, and approaches zero at $y \sim 0.30$. This composition range, which is richer in Cu than Cu$_2$Te, will be referred to as region I. The NMR study shows that K_{Cu} decreases rapidly with y in this region, as seen in Fig. 8.15.

In the composition range between Cu$_2$Te and CuTe (region II), the behavior is very unusual. The quantity σ rises rapidly from a minimum ~ 600 ohm^{-1} cm^{-1} at $y = \frac{1}{3}$ to ~ 2000 ohm^{-1} cm^{-1} at $y \cong 0.5$ and $d\sigma/dT$ is negative for $0.4 < y < 0.5$, as in a typical metal. Corresponding to this,

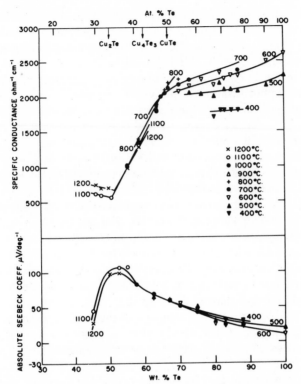

Fig. 8.14. The electrical conductivity and thermopower of the molten Cu–Te system (Dancy, 1965).

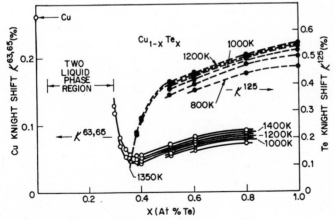

Fig. 8.15. Isotherms of the Knight shifts for ^{63}Cu and ^{65}Cu (open circles) and ^{125}Te (closed circles) versus composition for $Cu_{1-x}Te_x$. (Warren, 1973).

S decreases and has a positive temperature coefficient (again like a typical metal) in the range where $d\sigma/dT < 0$. In this region, K_{Cu} goes through a shallow minimum near $y \cong 0.38$ and then increases slowly with y, while K_{Te} increases rapidly with y for $y > 0.38$.

In the composition range $y > 0.5$ (region III), σ does not change much at constant T, but it does exhibit again a positive temperature coefficient corresponding to an activation energy ~ 0.1 eV. The quantity S decreases with y and dS/dT becomes positive and larger as $y \to 1$. Both K_{Cu} and K_{Te} increase slowly with y and have positive temperature coefficients.

Thus the overall behavior of Cu–Te is qualitatively different from Ag–Te and Tl–Te. Cu–Te has M-type behavior in region II, and S-type behavior in regions I and III, whereas the other two alloys have M-type behavior in region I and S-type behavior in regions II and III, with no apparent changes in behavior between II and III. The elements Cu, Ag, and Tl are chemically similar. Cu and Ag are noble metals (IB), whereas Tl is in group IIIB. But as noted earlier, Tl commonly exhibits group IB type of behavior because of the uncommonly low energy of the valence 6s electron shell. But Cu differs from Ag (and Tl) in frequently forming bivalent compounds. This reflects the fact that its filled 3d shell has a particularly high energy (compared to Ag and Au), so that one of the d electrons may be used for bonding in addition to the usual s electron. Therefore we suppose that a change in chemical bonding occurs, as y increases from $\frac{1}{3}$ to $\frac{1}{2}$, which involves the d electrons of the Cu atom. This suggests that the d electrons have a relatively high energy in Cu_2Te, and contribute strongly to the valence band at that composition. This forms the basis of the following tentative explanation of observed behavior.

In region I, we suppose that the conduction and valence bands are strongly overlapping. This is not unexpected, considering the very high temperature ($>1125°C$), since the band gap for Tl_2Te derived in Section 7.1 is ~ -0.45 eV at that temperature. In the present case, however, we suppose that the density of states in the valence band (N_v) rises much more rapidly than it does in the conduction band (N_c), probably as the result of the Cu 3d states, so that $N_v(E_f) > N_c(E_f)$ at the composition Cu_2Te, as shown in Fig. 8.16.

Fig. 8.16. Rigid band model for the electronic structure of $Cu_{1-y}Te_y$ at compositions near $y = \frac{1}{3}$. The overlap between the valence and conduction bands increases with temperature, and the Fermi energy decreases with y as indicated.

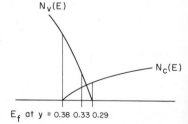

$N_V(E)$

$N_c(E)$

E_f at y = 0.38 0.33 0.29

This causes S to be positive. The large temperature coefficients of S and σ are explained by a large negative value of dE_G/dT. The effect of changing composition is then consistent with a rigid band model. As indicated in Fig. 8.16, adding Cu to Cu_2Te increases E_f, causing S to decrease as $N_c(E_f)$ increases and becomes comparable to $N_v(E_f)$. As noted earlier, the valence band states are expected to have largely d character, but the conduction band can be expected to have s character. This explains why K_{Cu} decreases rapidly with y. The positive value of dK_{Cu}/dT is also consistent with the expected increase in $N_c(E_f)$ with T due to the decrease in E_G.

As y becomes greater than $\frac{1}{3}$, the rigid band model suggests that E_f drops into the valence band, and $N_c(E_f)$ decreases to zero. The quantity K_{Cu} has a minimum near $y \cong 0.38$, and dS/dT and $d\sigma/dT$ become zero at this point, which suggests that the $N_c(E_f) \cong 0$ at this composition. At the same time, σ is increasing rapidly, and reaches a maximum at $y \cong 0.5$.

The different behavior of K_{Cu} and K_{Te} in region II is significant. The dependence of the Knight shift on $N(E_f)$ in Eq. 6.35 is the same for both nuclei. Therefore the fact that K_{Te} increases with y more rapidly than K_{Cu} suggests that the s-component of the wave functions (at E_f) is increasing for Te. We think that this is the result of an increasing admixture of the higher Te (6s) atomic orbital due to an increasing electrostatic potential at the Te atoms. This can be caused by the Madelung potential of the copper ions if the CuTe is ionic, or it may be caused by a positive charge on the Te atoms if there is three-fold bonding. As noted earlier, both Warren and Enderby have inferred that the bonding in CuTe is ionic, and Warren has found evidence that there is greater covalency in Cu_2Te. This seems consistent with a model in which CuTe is a mixture of $Cu_2Te + Te^{2-}$, and the positive charge provided is by holes in states of the valence band which are derived from Cu 3d orbitals.

To understand the behavior in region III, it should be recognized that a homogeneous liquid phase occurs only at $T \gtrsim 650°C$ for $y = 0.5$, but it extends to lower temperatures ($\gtrsim 400°C$) at larger values of y. At $T \gtrsim 600°C$, molten Te is fairly metallic, and has a three-fold bonding structure. The effect of changing composition at constant temperature from CuTe to Te can therefore be regarded as a homogeneous change between two metals with similar values of σ ($\sim 2000\ ohm^{-1}\ cm^{-1}$). As the Te concentration is increased, it is possible for σ to decrease with decreasing T for the same reason that it does so in pure Te (see Section 7.6). As far as it can be ascertained, K_{Cu} and K_{Te} both vary approximately as $\sigma^{1/2}$, indicating that the common cause of dK/dT and $d\sigma/dT$ is a decrease in $N(E_f)$. The σ and S data in this range has been examined for evidence of rigid band behavior by plotting $\ln \sigma$ versus $\ln(S/T)$. It is found that both Dancy's and our own data in the composition range $0.6 < y < 0.8$ fall on a common curve as is

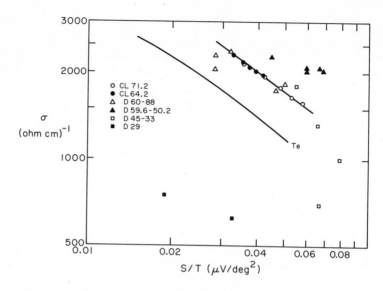

Fig. 8.17. The quantity ln σ versus ln(S/T) for $Cu_{1-y}Te_y$. Compositions are given in at% Te. A line is drawn through data at 60–80 at% Te, and this is compared with the curve for pure Te. Sources of data are D (Dancy, 1965) and CL (Cutler, 1963).

shown in Fig. 8.17. This curve is parallel to the one for pure Te, with a slope corresponding to $\sigma(E) \propto E^{0.7}$, but lies higher by a factor 1.4. For $y < 0.6$, the points differ widely from a common curve, indicating that the valence band structure is changing strongly.

8.5 In–Te AND Ga–Te ALLOYS

Thermoelectric measurements reported for In–Te (Ninomiya *et al.*, 1973; Popp *et al.*, 1974; Glazov *et al.*, 1969; Blakeway, 1969) and Ga–Te (Valient and Faber, 1974; Glazov *et al.*, 1969) show a pattern which is distinctly different from Tl–Te. The σ isotherms for In_xTe_{1-x}, shown in Fig. 2.6, dip at the composition In_2Te_3 rather than In_2Te. Very similar isotherms are reported by Valient and Faber for Ga–Te. We show in Fig. 8.18 the thermopower isotherms for In–Te of Ninomiya *et al.* (1973). They are similar to Tl–Te in that S changes sign near the composition In_2Te, but the magnitude

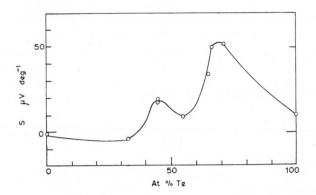

Fig. 8.18.   Dependence of the thermopower of In_xTe_{1-x} at 700°C on composition at%
Te. (Nakamura and Shimoji, 1973.)

of S on either side of the inversion is relatively small ($\gtrsim 20 \ \mu V/deg$). This
small magnitude is consistent with the large value of σ ($\sim 3000 \ ohm^{-1} \ cm^{-1}$).
Near the composition In_2Te_3, S has a maximum value $\sim 50 \ \mu V/deg$ at 700°C,
while $\sigma \sim 100 \ ohm^{-1} \ cm^{-1}$. The function $S(x)$ has a peculiar second peak
at ~ 60 at% In which has been found also by Popp *et al.* (1974). The $S(x)$
isotherm reported by Valient and Faber (1974) for Ga–Te is very similar
except that there is no local minimum between Ga_2Te and Ga_2Te_3. In
addition to these thermoelectric studies, Warren has made NMR measurements of In_2Te_3, Ga_2Te_3, Ga_xTe_{1-x}, and $Ga_2(Se_xTe_{1-x})_3$ (Warren, 1971,
1972b, 1973a).

   More detailed examination of the transport data shows further differences
from Tl–Te. There is no evidence of rigid band behavior since plots of $\ln(S/T)$
versus $\ln \sigma$ yield a different curve for each composition.[†] Also, in contrast
with many other chalcogenide alloys, plots of $\ln \sigma$ versus T^{-1} in the range
$x < \frac{2}{3}$ have slopes which vary both with composition and temperature. The
only similarity to Te–rich Tl–Te is that the activation energy is large in a
range $\sigma \gtrsim 2000 \ ohm^{-1} \ cm^{-1}$, which indicates that the responsible mechanism is a chemical reaction rather than purely electronic excitations.

   [†] Our investigation of this and other questions discussed in this section was aided by the
use of unpublished data for $\sigma(T, x)$ of Ga_xTe_{1-x} and In_xTe_{1-x} from the thesis of Lee (1971),
which was kindly made available by B. Lichter.

To understand the implications for the electronic structure, which we assume essentially the same for Ga–Te and In–Te, it is useful to first make a comparison with a superposed band model for ionic constituents, as proposed by Roth. We show schematically in Fig. 8.19 the s and p bands associated with the In^{+3} ions, and the p band due to the Te^{2-} ions. The total number of states is proportional to x for the In bands, and to $1 - x$ for the Te band. The relative positions of the In and Te bands are affected by charge transfer, but since ionic bonding is assumed, the Te band remains largely below E_f. Consequently, electrons are removed from the In bands as $1 - x$ increases, and there are just enough electrons left to fill the In s band at $x = \frac{2}{3}$. The change of sign of S at $x \cong \frac{2}{3}$ is consistent with E_f moving down through the region where the s and p bands overlap, as indicated in Fig. 8.19, and the magnitudes of σ and S indicate a strong overlap. (In_2Te may possibly be covalently bonded, but in view of the discussion of Section 8.1.2, the electronic structure would be similar.)

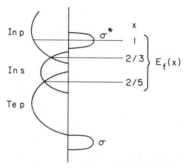

Fig. 8.19.   Model for the electronic structure of In_xTe_{1-x} in the range $x > 0.4$. Addition of Te removes electrons from the In p band for $x > \frac{2}{3}$. Between $x = \frac{2}{3}$ and $x = \frac{2}{5}$, covalent In–Te bonds are formed at low temperature, converting In s and Te p states into σ and σ^* states. At high T, these bonds are broken, causing the σ and σ^* states to return to their original bands. The dependence of E_f on composition is shown on the right.

As x decreases below $\frac{2}{3}$, the ionic model predicts that E_f will move down in the In s band, and if there were no overlap with the Te p band, the s band would be emptied when x reaches $\frac{2}{5}$. Although the total number of states in the s band decreases as x does, one would expect $N(E_f)$ to increase and then decrease. Since $\sigma < 3000$ ohm^{-1} cm^{-1}, we expect the behavior of S and σ to be described by diffusive transport (Eq. 6.14a) in the metallic approximation (Eq. 6.4). The qualitative behavior of $S(x)$ in this composition range is in accord with the expected decrease of E_f in a nearly rigid In s band. It rises as E_f leaves the region of overlap with the In p band, goes through a maximum at $x \cong 0.55$, and then a minimum at $x \sim 0.45$, corresponding to E_f moving down in the s band to a point where $(dN(E)/dE)_{E_f}$ becomes small. The rise in S as x drops below 0.4 corresponds to E_f moving into the larger, overlapping Te p band. However, this model is inconsistent with the observed behavior of σ. For one thing, σ decreases monotonically between $x = \frac{2}{3}$ and $\frac{2}{5}$, whereas the model suggests that σ should increase or at least decrease very slowly in the range $x \gtrsim 0.5$. Second, it is very difficult to explain the increased temperature dependence of S and σ as x decreases from $\frac{2}{3}$ to $\frac{2}{5}$ in terms of an ionic model.

These two factors suggest that $N(E_f)$ is decreasing more rapidly with decreasing x than predicted by the ionic model, and that the responsible mechanism is reversed on increasing T. We propose that the mechanism is the formation of covalent bonds between In and Te when $x < \frac{2}{3}$. As indicated in Fig. 8.5, the formation of covalent rather than ionic bonds causes states from the s band to rise into a higher antibonding σ^* band, and states from the Te p band to drop to a lower bonding σ band, and these additional bands are indicated in Fig. 8.19. (Of course, the σ^* and σ states are mixtures of In s and Te p states, in accord with Eq. 8.3.) The covalent bonds are much more rigid than ionic bonds, and lead to the formation of relatively large molecular clusters, which reduces the configurational entropy. Therefore an increase in T can be expected to reverse the process. With covalent bonding at low T, $N(E_f)$ will decrease rapidly as x decreases, since the total number of states in the s band is equal to four times the number of In_2Te molecules in the pseudobinary mixture $In_2Te–In_2Te_3$, i.e., $(5x - 2)$, rather than $2x$.

According to this model, a fraction $f(T)$ of the In_2Te_3 molecules in the pseudobinary mixture are ionically bonded, and the remaining In_2Te_3 molecules are covalent. This leads to a total number of s states (per atom) equal to $c_0 = (5x - 2) + (2 - 3x)f(T)$, and the number of electrons equal to $c = 5x - 2$. The shape of the s band is not known nor easily predictable, but one can make a rough assumption that σ is proportional to the hole concentration $c_0 - c = (2 - 3x)f(T)$. In fact, a very simple behavior is observed if one plots σ linearly versus T instead of the customary $\ln \sigma$ versus T^{-1} as shown in Fig. 8.20. In the metallic range of σ, the curves are described by:

$$\sigma = \sigma_1(T - T_0). \tag{8.8}$$

According to this crude model, the slope σ_1 should be proportional to $2 - 3x$ for $\frac{2}{5} < x < \frac{2}{3}$, and this seems to agree with the observed dependence of σ_1 on x, shown in Fig. 8.21. As shown also in the figure, T_0 increases rapidly to a maximum value at $x \cong 0.4$. The value of T_0 implies a critical

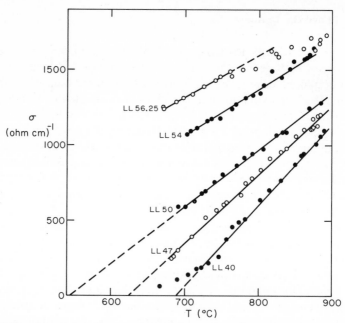

Fig. 8.20. Linear plots of σ versus T for In_xTe_{1-x} alloys in the composition range between 40 and 67 at% In. The data are from Lee (1971), and the compositions are given in at% In.

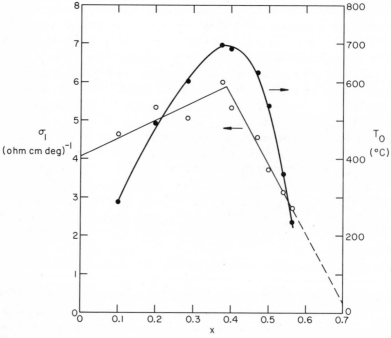

Fig. 8.21. Plots of the parameters σ_1 (○) and T_0 (●) defined in Eq. 8.8, as a function of composition in In_xTe_{1-x}.

temperature below which covalent bonding is complete. The function $\sigma(T)$ for $x < \frac{2}{5}$, not shown in Fig. 8.20, is also in accord with Eq. 8.8, and Fig. 8.21 shows that σ_1 decreases moderately as x goes to zero, and T_0 falls off strongly. This suggests that the model can be extended to the pseudobinary mixture $In_2Te_3 + Te$. The Te constituent is not fully ionic, in rough accord with the picture arrived at in Section 7.6 for molten Te. It is not clear why $f(T)$ should be linear in T rather than follow the usual Arrhenius formula. We suspect that the reason may be related to the strong effect of the configurational entropy, caused by the formation of large clusters, on the bond equilibrium.

We turn now to the results of NMR studies in Ga_xTe_{1-x} (Warren, 1972b). The isotherms of the Ga Knight shift K_{Ga} for the various temperatures are shown in Fig. 8.22. Warren also reported isotherms for K_{Te} and the magnetic R_M and quadrupole R_Q relaxation rates for Ga nuclei. Both relaxation rates peak near $x = 0.4$, and have a large negative temperature coefficient. Warren points out that the peak in R_M reflects a large enhancement over the Korringa rate (see Section 6.4.3) due to partial localization which occurs when $N(E_f)$ is small, and the peak in R_Q indicates the presence of long-lived molecular clusters. (He calculates 10^{-11} sec.) The quantity R_Q increases by a factor ~ 3 between 1300 and 1100°K, indicating a rapid decrease in cluster lifetime

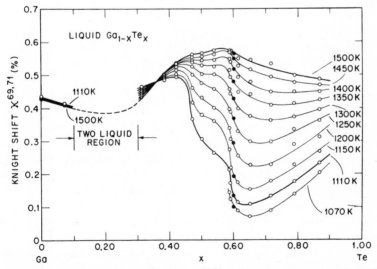

Fig. 8.22.   Isotherms of Ga Knight shifts $K^{69,71}$ as functions of concentration in liquid $Ga_{1-x}Te_x$. The closed circles denote data obtained at Ga_2Te_3. The dashed line represents an extrapolation across the miscibility gap. At other compositions, the isotherm for 1110°K is the lowest one which lies above the liquidus temperature (Warren, 1972).

with T. All of this is in accord with the inferences which we made from the behavior of S and σ.

Let us consider the behavior of K_{Ga} in more detail, assuming that its value reflects the s component of $N(E_f)$ for wave functions at Ga sites. At 31% Te, where we think E_f is somewhat above the minimum in $N(E)$ due to overlapping s and p bands (Fig. 8.19), dK/dT is negative, in contrast to the behavior at larger values of x. This implies that the s band contribution to $N(E_f)$ decreases with T, and it might be caused by band narrowing due to thermal expansion. In the range $0.5 < x < 0.6$, we find that plots of $\ln K$ versus $\ln \sigma$ are linear with a slope corresponding to Eq. 6.33. This is in agreement with the expectation that the s character of the wave functions at E_f at Ga nuclei will be unchanged when the covalent bonds dissociate. Similar plots for Te–rich compositions $x < 0.40$ are no longer in accord with Eq. 6.33. The quantity E_f is now in the Te p band, and this suggests that the s electrons due to the ionized In atoms form a band which still strongly overlaps this band, and this changes the s contribution to $N(E_f)$. If we assume that $N(E_f)$ is the sum of p contributions N_p and s contributions N_s, and substitute K/k for $N_s(E_f)$, where k is a constant, the following relation is derived from Eq. 6.14a:

$$\sigma^{1/2} = A^{1/2}[N_p(E_f) + K/k]. \tag{8.9}$$

Plots of K_{Ga} versus $\sigma^{1/2}$ are shown in Fig. 8.23 for a number of compositions, and it is seen that they are in accord with Eq. 8.9. The value $\sigma_0 \sim A[N_p(E_f)]^2$ at which the lines intersect $K = 0$ are nearly equal to zero for $x \gtrsim 0.4$. From $x = 0.35$ to 0.17, σ_0 increases. The slopes $dK/d\sigma^{1/2}$ are nearly the same (within 10%) for all compositions except $x = 0.17$. The quantity $dK/d\sigma^{1/2}$ is ~ 2 times larger for $x = 0.17$, and the reason for this is not apparent. These results seem to support the validity of the dissociating covalent bond model.

Warren (1973b) has reported an interesting study of σ and NMR behavior in the pseudobinary alloy $Ga_2(Se_xTe_{1-x})_3$. He found that the magnetic relaxation enhancement factor η_R has a dependence on the Knight shift K_{Ga} that is independent of whether T or x is varied over a wide range of both variables. The experimental range of σ extends above and below the value at the mobility edge $\sigma(E_c)$ (~ 200 ohm^{-1} cm^{-1}) by large factors. Of course, both η_R and σ represent averages of processes at energies above and below E_f in nonmetallic situations, but Warren was able to obtain a self-consistent interpretation of the data by assuming that the relationship between η_R and $\sigma(E_f)$ in Eq. 6.36 remains valid when E_f is below the mobility edge. Using this, together with the empirical curve for η_R versus K_{Ga} and the proportionality of the Knight shift to the density of states (Eqs. 6.30 and 6.33), he finds that $\sigma(E_f) \propto [N(E_f)]^{2.6}$. He also derives theoretical curves

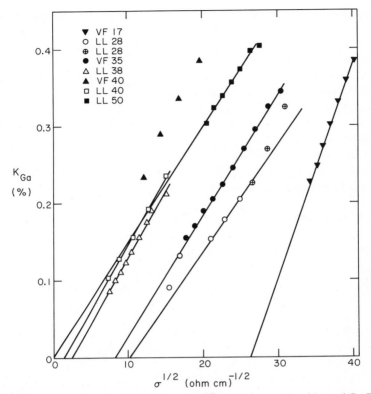

Fig. 8.23.   The gallium Knight shift versus $\sigma^{1/2}$ for various compositions of Ga_xTe_{1-x}, given in at% Ga. The Knight shift data are from Warren (1972b) and the sources of the conductivity data are VF (Valient and Faber, 1974) and LL (Lee, 1971). In some cases, the compositions for K and σ differ by one or two at% Ga.

for $\sigma(T)$ at the various compositions which are in very good agreement with experiment, using a model shape for $N(E)$:

$$g(E) = N(E)/N^0(E) = \tfrac{1}{2}[(1 + g_m) + (1 - g_m)\cos(\pi E/\Delta E)], \quad (8.10)$$

where $N^0(E)$ is the free-electron density of states; g_m is the value of g at the center of the pseudogap, and it is inferred from K_{Ga}. ΔE is the width of the pseudogap, and its value is chosen to yield the correct σ at the lowest T for each composition.

The resulting curves for $\sigma(E)$ rise steeply in the neighborhood of $\sigma(E_c)$, but $d\sigma/dE$ is much smaller than what one would expect for a discontinuity of a factor of 10–100 smoothed out over an energy interval $\sim kT$. From this, Warren concludes that the mobility edge is softer than what is indicated by the theories of Mott and Cohen. It is to be noted that $\sigma(E)$ is inferred from $N(E)$ at energies other than the center of the pseudogap, using an empirical relationship obtained from situations in which $E\ (=E_f)$ is at the center of the pseudogap. The averaging processes in σ and η_R for energies above and below E_f when E_f is near the mobility edge are different for the two situations, so that the conclusion about the softness of the mobility edge is based on rather indirect evidence. The interpretation of η_R when E_f is near or below the mobility edge is a question that seems to warrant further study, and of course other experimental tests of the nature of the mobility edge in liquids are clearly desirable.

8.6   Se–Te ALLOYS

8.6.1   Pure Selenium

We come now to the less commonly studied category of high-resistivity liquids. Pure selenium has been studied for a long time. Compared to other semiconducting liquids, its molecular structure is well understood. Liquid sulfur is very similar, and both liquids have been shown to consist of a mixture of chain molecules and 8-membered rings (and perhaps larger ones). The concentration of rings increases with decreasing T, and there is a critical temperature below which the only ring molecules occur. The critical temperature is observed experimentally only in sulfur; its calculated value is below the freezing point for selenium. The theory of bond equilibrium for these liquids is very well established (Eisenberg and Tobolsky, 1960; Gee, 1952), and it is the prototype for the theory described in Section 7.3 for Tl–Te alloys. For the present discussion, it suffices to note that the average chain length decreases with T, and the concentration of dangling bonds is described in terms of an equilibrium constant with an activation energy E_d which has been determined in several ways. Eisenberg and Tobolsky (1960) estimated $E_d = 0.54$ eV on the basis of viscosity data. The dangling bonds are paramagnetic centers, and determination of their concentration versus T by electron spin resonance has yielded $E_d = 0.63$ eV (Koningsberger et al., 1971). A value $E_d = 0.87$ eV has resulted from a study of the magnetic susceptibility (Massen et al., 1964).

Information about the electronic structure has been obtained in studies of the electrical conductivity (Lizell, 1952; Baker, 1968; Gobrecht et al., 1971a), the thermopower (Gobrecht et al., 1971b), and the optical absorption spectrum (Perron, 1972). Perron derived a band gap (determined for a constant optical absorption coefficient $\alpha = 1000$ cm^{-1}) which can be expressed as $E_K = E_{K0} - E_{K1}T$, with $E_{K0} = 2.20$ eV and $E_{K1} = 1.23 \times 10^{-3}$ eV/deg. Three ranges of temperature can be distinguished in which $\sigma(T)$ behaves differently. In the lowest range, where $\sigma \gtrsim 10^{-5}$ ohm^{-1} cm^{-1},

σ is sensitive to impurities, particularly oxygen, and these results can be neglected here as reflecting extrinsic behavior. In an intermediate range $(770 < T < 990°\mathrm{K})$, behavior characterizing the pure liquid is observed, and the activation energy $E_{\sigma 1}$ determined in various measurements has a spread of values. Gobrecht and coworkers (1971a) in their study seem to have achieved the greatest control of impurities; their value of $E_{\sigma 1}$ is 1.25 eV. At $T > 990°\mathrm{K}$, a higher activation energy is observed, whose value is more poorly determined: $E_{\sigma 2} = 2.0 \pm 0.3$ eV (Gobrecht *et al.*, 1971a). Considerable light is shed on the situation by a study of the thermopower. The curve for $S(T)$ from the work of Gobrecht *et al.* (1971b) is shown in Fig. 8.24.[†] The quantity S is negative at low T, but it rises sharply above 770°K and becomes positive for $T > 990°\mathrm{K}$. The activation energy dS/dT^{-1} in the range between 770 and 990°K is much larger than $E_{\sigma 1}$, and we think that the large slope reflects ambipolar transport. The most straightforward interpretation of the thermoelectric data seems to be that electron transport, possibly due to impurities, is dominant at $T < 990°\mathrm{K}$ and the generation of holes by bond breaking leads to the activation energy $E_{\sigma 1}$. At $T > 990°\mathrm{K}$, this is superseded by excitation of electron-hole pairs across a band gap, and since the holes are more mobile, S is positive. The quantity $E_{\sigma 2}$ then is the activation energy for electron-hole pairs. The difficulty with this interpretation is that it requires a band gap $\sim 2E_{\sigma 2} \cong 4.0$ eV. (We are referring, of course, to the extrapolated value at $T = 0°\mathrm{K}$.) This conflicts with the optical data, which indicates that

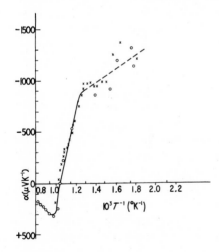

Fig. 8.24. Thermoelectric power of pure liquid selenium as a function of reciprocal temperature (Gobrecht *et al.*, 1971b).

[†]A later report (Mahdjuri, 1975) makes it clear that there was an error in the sign of S and this has been corrected in Fig. 8.24.

the band gap is $E_{K0} = 2.0$ eV. We think that $E_{\sigma 2}$ is not a real activation energy, and that the larger value of $d \ln \sigma/dT^{-1}$ reflects a merging of bands. This causes $\sigma(E_1)$ to increase rapidly with T, where E_1 is the energy at which transport occurs. The proposed process is discussed in more detail below in relation to Te–Se alloys, where the same phenomenon apparently occurs.

The boiling point of Se at atmospheric pressure is 958°K, and the experiments of Gobrecht *et al.* were done in pressurized apparatus. Very high pressure measurements have been reported recently (Hoshino *et al.*, 1976) which go into the supercritical range of temperatures and pressures. An interesting aspect of these results, shown in Fig. 8.25, is that $\sigma(T)$ is qualitatively similar to Se–Te alloys and pure Te (Fig. 2.5) in that σ levels off in the semimetallic range ($\sigma > 100$ ohm^{-1} cm^{-1}). The authors point out that the density decreases strongly with T in this range, and the drop in σ at the highest T reflects an increasing separation of the atoms.

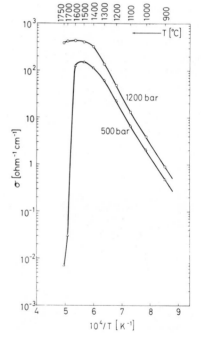

Fig. 8.25. Electrical conductivity isobars for fluid selenium in the supercritical range (Hoshino *et al.*, 1976).

8.6.2   Se–Te

Information about the molecular structure is to be found in studies of the viscosity (Ioffe and Regel, 1960) and NMR (Seymour and Brown, 1973). As discussed in Section 3.3, the dependence of η on T and x in Se$_x$Te$_{1-x}$ indicates that the average length of the molecules decreases with T and increasing tellurium content. This implies that the Te–Te bonds are weaker than the Se–Se bonds. Although the NMR relaxation times that were measured in pure selenium are in good accord with the bond lifetimes inferred from the average chain length, the information for the alloys is more limited. Seymour and Brown made an attempt to infer from their results whether Te–Se bonding is preferred over random bonding, with ambiguous results. A preference would indicate that the Te–Se bond energy is greater than the average for Se–Se and Te–Te bonds. Such a difference ΔE_{TS} is expected to the extent that there is ionicity in the bond (Pauling, 1960). Evidence that $\Delta E_{TS} \gtrsim 0.07$ eV has been obtained in a study of the Mössbauer effect in disordered crystalline $Se_x Te_{1-x}$ alloys (Boolchand and Suranyi, 1973).

The thermoelectric properties of $Se_x Te_{1-x}$ alloys have been thoroughly studied in the Te-rich range (Perron, 1967). The curves for ln σ and S versus T^{-1} (Figs. 2.5 and 2.9) pose an intriguing problem in the range of lower conductivities ($\sigma \ll 200$ ohm^{-1} cm^{-1}) where the Maxwell–Boltzmann approximation should be valid. According to Eqs. 6.5, ln σ and S should have the same activation energy. We list the activation energies in Table 8.3. The σ curves have two activation energies for $x = 0.3$ to 0.5. The larger one, labeled $E_{\sigma 2}$, occurs in a region (II), where $\sigma \gtrsim 1$ ohm^{-1} cm^{-1}, and it has values ~ 1.3 eV. The smaller one, $E_{\sigma 1}$, in region I, is well defined only for $x = 0.5$, and it has values ~ 0.8 eV. The curves for S versus T^{-1} span both

TABLE 8.3

Activation Energies of $Se_x Te_{1-x}$

x	$E_{\sigma 1}$ (eV)	$E_{\sigma 2}$[a] (eV)	E_{S0}[a] (eV)	E_{S1} (10^{-3} eV/deg)	E_{K0}[a] (eV)	E_{K1}[a] (10^{-3} eV/deg)
0	—	0.52	0.13	—	—	—
0.05	—	0.63	0.58	—	—	—
0.1	—	1.19	0.85	—	—	—
0.2	—	1.27	0.83	1.11	—	—
0.3	~ 0.8	1.28	0.87	0.98	—	—
0.4	~ 0.8	1.35	0.86	0.83	—	—
0.5	0.80	1.37	$(1.19)^b$	0.72	—	—
0.6	—	—	—	—	$(1.5)^b$	$(1.1)^b$
0.7	—	—	—	—	1.7	1.15
0.8	—	—	—	—	1.97	1.27
0.9	—	—	—	—	2.02	1.14
0.95	—	—	—	—	2.16	1.19
1.00	1.25	~ 2.0	—	—	2.20	1.23

[a] Numbers taken from Perron (1972) except for $E_{\sigma 2}$ at $x = 1.0$.
[b] Doubtful numbers are enclosed in parentheses.

regions with constant slope $E_{S0} \sim 0.8$ eV, except for $x = 0.5$, where it is appreciably larger. Thus E_{S0} agrees well with $E_{\sigma 1}$ in most cases, but not with $E_{\sigma 2}$.

Mott and Davis (1971) have considered this problem and came to the conclusion that $E_{\sigma 2}$ is not a true activation energy, but reflects a rapid increase in the conductivity near the mobility edge with T. We agree with this interpretation, which suggests that Eqs. 6.5 are valid with constant values of $\sigma(E_1)$ only in region I. It is interesting to determine the apparent value of $\sigma(E_1)$ from the experimental values of σ and S, using Eqs. 6.5. When this is done for $x = 0.3$–0.5, the values are found to lie in a range 40–120 ohm^{-1} cm^{-1}. The function $\sigma(E_1)$ is independent of T for $x = 0.4$ and 0.3. In the case of $x = 0.5$, $\sigma(E_1)$ decreases with T, reflecting the difference in values between $E_{\sigma 1}$ and E_{S0}. The magnitudes of $\sigma(E_1)$ correspond to hole transport at a mobility edge, but they are somewhat smaller than the expected value (~ 200 ohm^{-1} cm^{-1}). The apparent variation of $\sigma(E_1)$ with T for $x = 0.5$ may be explained if S contains an electron contribution which is increasing with T, and the relatively small values of $\sigma(E_1)$ at the other compositions suggest also an ambipolar effect, this time one which is constant with T. Assuming that this is the case, Eq. 6.25 gives:

$$\Delta S = S_p - S = (S_p - S_n)\sigma_n/\sigma. \qquad (8.11)$$

We have determined ΔS, on the assumption that the conductivity at the mobility edge is 200 ohm^{-1} cm^{-1}. Equations 6.5 and 8.11 then give:

$$\Delta S = (k/e) \ln[200/\sigma(E_1)]. \qquad (8.12)$$

The magnitudes of ΔS are ~ 100 μV/deg, and plots of S_p versus T^{-1} now are parallel for $x = 0.2$ to 0.5, with the same slope $E_{S0} \cong 0.85$ eV. Designating the energy at the mobility edge now as E_{v1}, we write $E_f - E_{v1} = E_{S0} - E_{S1} T$, and putting this into Eq. 6.5b gives:

$$S_p = (k/e)[1 - (E_{S1}/k) + E_{S0}/kT]. \qquad (8.13)$$

Comparison with the experimentally derived curves gives values of E_{S1} as a function of x which are listed in Table 8.3.

These results show that a relatively small correction for an ambipolar contribution causes $S(T)$ and $\sigma(T)$ to conform to transport at a mobility edge with a constant activation energy in region I, and there is a systematic shift in E_{S1} with composition which will be discussed later. Since the $S(T)$ curves extend into region II without a break, it is reasonable to assume that S_p given in Eq. 8.13 describes the behavior of $E_f - E_{v1}$ also in region II. This provides a means of determining the excess conductivity $\Delta\sigma = \sigma - \sigma_p$ as a function of $E_f - E_{v1}$, where it is assumed that:

$$\sigma_p = 200 \exp[-(E_{S0} - E_{S1} T)/kT]. \qquad (8.14)$$

The quantity $\Delta\sigma$ is plotted versus $(E_f - E_{v1})/kT$ in Fig. 8.26, and σ_p is shown for comparison. It is to be noted that $\Delta\sigma$ rises much more rapidly than σ_p and exceeds it when $E_f - E_{v1} \gtrsim 2kT$. A plausible explanation for $\Delta\sigma$ is that it is caused by overlap of a higher electron band with the valence band. The quantity E_f can be expected to lie approximately half way between the mobility edges E_{v1} and E_{c1} in the two bands. That would account for the nearly constant ambipolar contribution ΔS at lower T, which implies that σ_n/σ_p is nearly constant with changing T. The results in Fig. 8.26 then imply that the mobility gap $\gtrsim 4kT$ at the point where $\Delta\sigma$ exceeds σ_p. This implies that there is a very deep tail (>0.3 eV) on the conduction band if $\Delta\sigma \sim \Delta\sigma_p$, or on the valence band if $\Delta\sigma \sim \Delta\sigma_n$. The former case is illustrated in Fig. 8.27a. Perron (1972) has proposed that there is a deep tail on the conduction band, on the basis of an analysis of optical absorption curves for alloys with compositions $x > 0.5$ (Perron, 1969). Another possibility is that the electron band is an acceptor impurity band caused by dangling bonds. In that case,

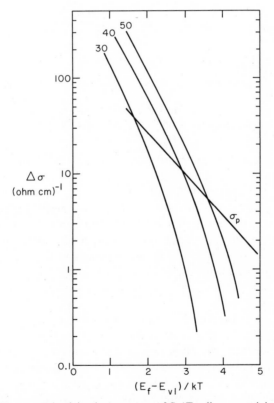

Fig. 8.26. Excess conductivity $\Delta\sigma = \sigma - \sigma_p$ of Se–Te alloys containing 30, 40, and 50 at% Se, as a function of $E_f - E_{v1}$ in units of kT.

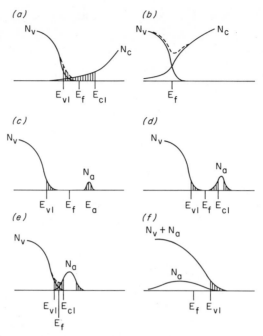

Fig. 8.27. Overlapping band models for Se–Te alloys. (a) and (b) are the low and high temperature limits when the second band is the conduction band. (c), (d), and (e) describe the growth of an acceptor band which is merged in the high temperature limit (f). The hatch marks indicate localized states, and the mobility edge is at E_{v1} for holes and at E_{c1} for electrons.

$\Delta\sigma_n$ can be expected to rise rapidly with temperature in both regions I and II because the total number of states in that band is increasing. As shown in Fig. 8.27c–e, the average value of $\sigma(E)$ in the acceptor band increases as extended states appear and extend their range of energy. In this case, a large part of $\Delta\sigma$ is due to $\Delta\sigma_n$. It should be noted, however, that these interpretations still have a difficulty in that they require a very special behavior in the thermopower S_1 associated with the extra transport. In order that no irregularity appear in $S(T)$ when $\Delta\sigma \gtrsim \sigma_p$, it is necessary that $S_1 \sim S$, so that the mean energy of the states contributing to $\Delta\sigma$ must continue to lie somewhat above the mobility edge.

The behavior of the thermopower in the metallic range ($\sigma \gg 200$ ohm^{-1} cm^{-1}) provides further information about the character of the electron band. Normally E_f lies near the minimum in the overall density of states when a conduction band overlaps the valence band as illustrated in Fig. 8.27b, and S drops to very low values in relation to σ. In such a case, a plot of $\ln(S/T)$ versus $\ln \sigma$ will have a very small slope, corresponding to small values of $d \ln \sigma/dE$ at E_f. This plot is shown in Fig. 8.28 for various compositions of

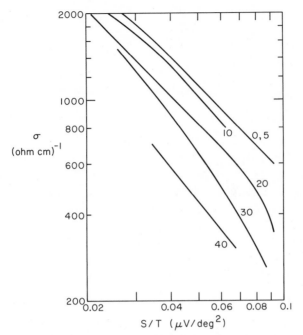

Fig. 8.28.   Plots of ln σ versus ln(S/T) for various compositions of Se–Te alloys, with compositions given in at% Se.

Se_xTe_{1-x}. It is seen that they are in accord with single-band transport, although the band is not rigid with changing x. Over the range of σ between 500 and 1500 ohm^{-1} cm^{-1}, $d \ln \sigma / d \ln(S/T) \sim -1$, which corresponds to $\sigma \propto E$. This supports the conclusion that the electron band is an acceptor band which is entirely absorbed into the valence band in the metallic range, as shown in Fig. 8.27f. Such a process is to be expected in view of the discussion in Section 7.3, since dangling bonds are expected to form acceptor states within the band gap at low hole densities, and these states are expected to be merged into the valence band at high hole densities.

Returning to the transport in the Maxwell–Boltzmann region, a striking aspect is the parallel behavior of ln σ and S versus $1/T$ for various compositions. It is reminiscent of the conductivity curves in Tl–Te, and one may ask whether it can be accounted for by a theory for bond equilibrium. The transport behavior reflects the temperature dependence of $E_f - E_{v1}$. If there is an acceptor band at an energy E_a, E_f is related to the density of acceptor states N_a and the hole density p by:

$$p = N_d / [1 + 2 \exp\{(E_a - E_f)/kT\}]$$
$$\cong (N_d/2) \exp[(E_f - E_a)/kT]. \tag{8.15}$$

One can also write:

$$p = N_v \exp[(E_{v1} - E_f)/kT], \qquad (8.16)$$

where N_v is a slowly varying function of T which depends on the shape of the valence band edge. Combining these equations gives:

$$E_f - E_{v1} = (E_a - E_{v1})/2 + (kT/2) \ln(2 N_v/N_d). \qquad (8.17)$$

The temperature dependence of N_d is governed by the equation for equilibrium of dangling bonds. In Appendix C3 an independent bond model is presented for Se–Te alloys. It depends on the dissociation energies of three kinds of bonds. There is evidence (Section 8.6.1) that the energy E_{SS} of Se–Se bonds is ~ 1.2 eV, and the study of bond equilibrium in Tl–Te (Section 7.3.3) indicates that $E_{TT} = 0.5$ eV for Te–Te bonds. We have noted evidence that the energy E_{TS} of a Te–Se bond exceeds the average of E_{SS} and E_{TT} by an amount ΔE_{TS} which is greater than kT. In that case, the parameter $k_m < \frac{1}{2}$ in Eq. C21, and the concentration b_{TS} of Se–Te bonds is described approximately by Eq. C28b at low T. Since $E_{SS} \gg E_{TT}$, the dangling bonds are primarily due to broken Te–Te bonds at low T, and their concentration is given by Eqs. C22a and C28b for $x < \frac{1}{2}$. Writing $K_{TT} = \exp[-2(E_d - TS_d)/kT]$ and $N_d = c_T N_a$ (N_a is the concentration of atoms), the result is:

$$N_d = (1 - 2x)^{1/2} N_a \exp(S_d/k) \exp(-E_d/kT). \qquad (8.18)$$

The result in Eqs. 8.17 and 8.18 can be compared with the experimental results for $E_f - E_{v1} = E_{S0} - E_{S1}T$. It is seen that:

$$E_{S0} = (E_a - E_{v1} + E_d)/2, \qquad (8.19)$$

and with $E_d = 0.25$ eV, the value $E_a - E_{v1} = 1.45$ eV is derived. This is in approximate agreement with the value of the optical gap E_{K0} obtained by extrapolating Perron's results (Table 8.3) to $x < 0.5$. The dangling bond model also predicts correctly that E_{S0} should be independent of x for $x < 0.5$, and that there is a change in behavior for $x > 0.5$ due to the disappearance of the weak Te–Te bonds.

The explicit dependence on x in Eq. 8.18 implies a change in E_{S1} with x which is an order of magnitude smaller than the experimental dependence shown in Table 8.3. This parallels the failure of the independent bond model to describe accurately the effect of composition on σ in Tl–Te, and similar explanations may be appropriate. However, there is evidence, discussed above, that the term $E_a - E_{v1}$ in Eq. 8.17 decreases rapidly with increasing T and decreasing x, and this is attributed to screening. Therefore it is reasonable to replace $E_a - E_{v1}$ by $E_{a0} - E_{a1}T$, so that $E_a - E_{v1}$ in Eq. 8.19

becomes the zero temperature limit E_{a0}, and the temperature coefficient E_{a1} appears in E_{S1}. We suggest that the strong dependence of E_{S1} on x is mainly due to the sensitivity of E_{a1} to composition.[†]

8.7 Tl–S AND Tl–Se ALLOYS

These systems are of special interest because of their chemical similarity to Tl–Te. Comprehensive thermoelectric measurements have been made in the range $0.4 \gtrsim x \gtrsim 0.7$ for both Tl_xSe_{1-x} (Nakamura and Shimoji, 1969; Regel et al., 1970) and Tl_xS_{1-x} (Nakamura et al., 1974; Kazandzhan and Tsurikov, 1974a). In addition the electrical conductivity of Tl_xSe_{1-x} has been investigated in the range $0 \gtrsim x \gtrsim 0.4$ (Petit and Camp, 1975). The isotherms for $\sigma(x)$ and $S(x)$ are shown in Fig. 8.29 for Tl_xSe_{1-x} and in Fig. 8.30 for Tl_xS_{1-x}. They are seen to be qualitatively similar to Tl_xTe_{1-x} (Fig. 2.3), the main difference being that the values of σ drop well below 100 ohm^{-1} cm^{-1} in the range $x \gtrsim \frac{2}{3}$. Corresponding to this, the magnitude of S becomes as large as 300 $\mu V/deg$. In addition, $\sigma(x)$ tends to be constant for $x < \frac{2}{3}$, extremely so in the sulfide system, whereas it increases with decreasing x in the telluride system.

The strong dependence of σ and S on x when $x > \frac{2}{3}$ suggests that the excess Tl is ionic as in Tl–Te. The effects of doping with a third element have been studied in both the sulfide and selenides (Nakamura and Shimoji, 1969; Nakamura et al., 1974; Kazandzhan and Tsurikov, 1973, 1974b; Kazandzhan et al., 1974). Using methods analogous to that described in Section 7.1 for Tl–Te (Cutler, 1973), the Japanese group has determined valences for Tl and a number of impurities which are listed in Table 7.1. A valence of one was found for Tl in the Se and S alloys as in the Te system, and the impurities differ only in the case of In, which is three in Se and S instead of one as in Te.

In contrast to Tl–Te, much of the data for $x > \frac{2}{3}$ lie in a range ($\sigma < 100$ ohm^{-1} cm^{-1}) where the MB approximation for σ and S can be made (Eq. 6.5). When $(x - \frac{2}{3})$ is very small, ambipolar effects may occur, or else the electron density may be affected by chemical dissociation of Tl_2Se or

---

[†] Since the completion of the manuscript, new work has appeared (Street and Mott, 1975; Mott et al., 1975) that makes it evident that the situation created by dangling bonds is a good deal more complicated than our description. It now seems probable that dangling bond atoms enter into threefold bonding configurations such as discussed in Section 7.6 to form donor states as well as acceptor states. These donor states may be close enough to E_f to seriously modify the results in Eqs. 8.15–8.19. Electrically neutral threefold bonding configurations containing a promoted electron in the σ^* state also form, and this causes a contribution to the energy of the acceptor states that has been missing in our discussions. The full implications of these new insights have yet to be worked out.

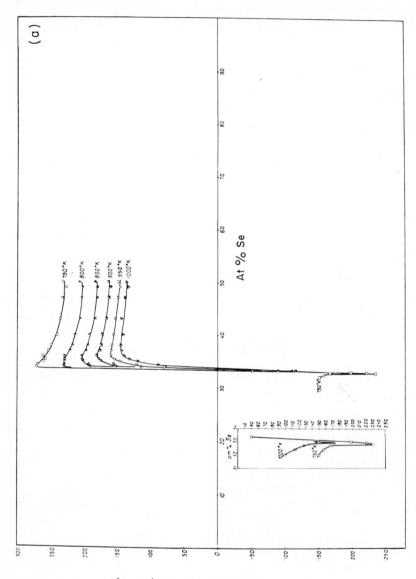

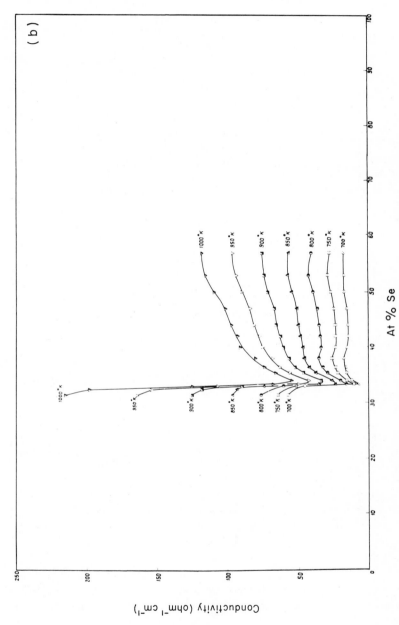

Fig. 8.29.   Isotherms of (a) S and (b) σ for Tl–Se alloys (Regel, *et al.*, 1970).

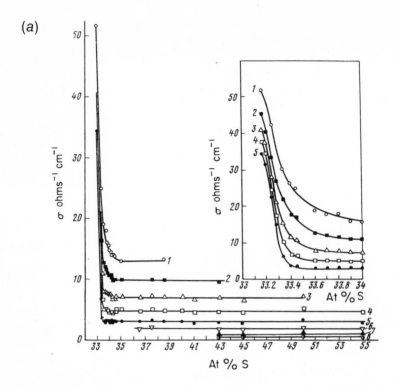

Fig. 8.30. Isotherms of (a) σ and (b) S for Tl–S alloys. Temperatures in °K are (1) 950, (2) 900, (3) 850, (4) 800, (5) 750, (6) 700, (7) 650, (8) 600 (Kazandzhan and Tsurikov, 1974c).

Tl_2S in addition to the concentration of excess Tl. The data of Nakamura, Matsumura, and Shimoji (NMS) for $\sigma(T)$ in Tl_2S + Tl, which is shown in Fig. 8.31, seems most amenable for examination, since it lies in the MB range and includes relatively large concentrations x_{Tl} of excess Tl (given in atomic percent). It is seen that there is a relatively small activation energy which increases as x_{Tl} decreases. This might be explained by a model in which E_f lies between the mobility edge E_{c1} and the band edge E_{c0} so that it changes very little with T. An alternative hypothesis, that n is increasing with T because of excitation of electrons from the valence band as in Tl–Te (Section 7.1), seems to be precluded by the larger band gap. Since the MB approximation has appreciable error, a more refined analysis was made using modified Fermi–Dirac integrals for the mobility edge model (Appendix B3). Assuming that the density of states is parabolic, $\sigma(x, T)$ is determined

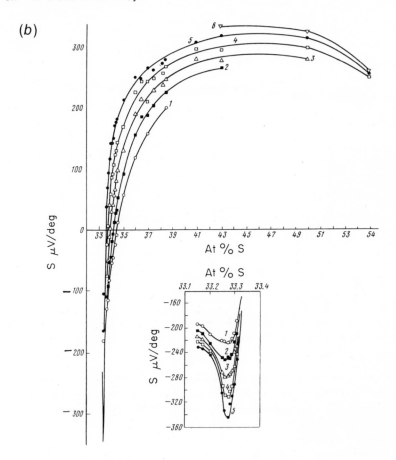

S μV/deg

At % S

At % S

S μV/deg

by Eqs. 7.1, 7.3, 7.7, and B7a. The theoretical curves shown in Fig. 8.31 were obtained with the following parameters (with the corresponding values for Tl_2Te given in parentheses): $AC^2 = 2700$ ohm^{-1} cm^{-1} eV^{-1} (2300), $C = 1.5 \times 10^{22}$ cm^{-3} eV$^{-3/2}$ (1.49×10^{22}), and $E_{c1} = 0.245$ eV ($\gtrsim 0.07$ eV). There is good agreement in the magnitude and slope of $\sigma(T)$ for compositions $x_{Tl} = 2$, 1.5, 1.0 and 0.5 at% excess Tl. The discrepancy in the magnitude of σ at 0.3 at% Tl may be due to chemical dissociation of Tl_2S. The parameters A and C agree very well with the ones obtained for Tl_2Te. There are however two disturbing aspects of the results. The energy of the mobility edge E_{c1} is rather large, and it indicates a value $\sigma(E_{c1}) = 660$ ohm^{-1} cm^{-1}, which is more than twice as large as expected. In addition, the model predicts values of the thermopower (Eq. B9b) which are larger than experiment by $\sim 150\mu V/\text{deg}$ at larger values of x_{Tl}. A possible reason

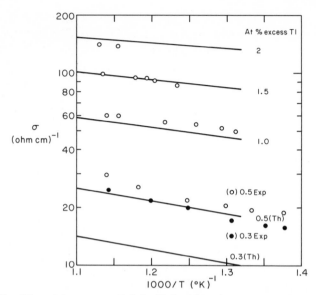

Fig. 8.31. Plots of ln σ versus T^{-1} for Tl_2S with various concentrations of excess Tl. The experimental points are from the data in Nakamura *et al.* (1974), and the theoretical curves are derived from a mobility edge model with a parabolic density of states.

for the discrepancies may be the assumption that $N(E)$ is parabolic. The band edge may well be distorted by band tailing. Further study is necessary to determine whether a different $N(E)$ would yield better agreement between experiment and theory.

Let us now examine the range $x < \frac{2}{3}$. In Tl_xSe_{1-x}, $\sigma(T)$ and $S(T)$ have the same activation energy for $x \cong 0.60$, in accord with Eqs. 6.5. Writing $E_1 - E_f$ as $E_{S0} - E_{S1}T$, where E_{S0} is the activation energy, the experimental curves yield $E_{S0} = 0.306$ eV, $E_{S1} = 2.50 \times 10^{-4}$ eV/deg, and $\sigma(E_1) = 161$ ohm^{-1} cm^{-1}. This indicates that transport is at a mobility edge rather than in localized states. The situation becomes less clear at smaller x. As x decreases, the activation energy for the conductivity E_σ increases, whereas the thermopower activation energy E_{S0} remains constant. It can be seen from Fig. 8.29b that the increase in E_σ occurs at higher temperatures where $\sigma \sim 100$ ohm^{-1} cm^{-1}, and E_σ remains constant at lower T.

In the case of Tl_xS_{1-x}, σ and E_σ are remarkably constant for $x < 0.65$. Ambipolar effects are visible in the thermopower for $x > 0.60$, $S(x)$ is nearly constant for lower concentrations of thallium. The difficulty here is that the thermopower has a lower activation energy $E_{S0} = 0.22$ eV, compared to $E_\sigma = 0.44$ eV in the results of Kazandzhan and Tsurikov (KT) shown in Fig. 8.30a. The results of NMS are somewhat different—$E_\sigma = 0.35$ eV—but still in poor agreement with the thermopower. A value of E_σ larger than E_{S0} can be expected if transport is in localized states, since $\sigma(E_1)$ in that case contains an activation energy $W = E_\sigma - E_{S0}$ due to hopping. But that interpretation leads to the inconsistent result that $\sigma(E_1) \cong 60$ ohm^{-1} cm^{-1} when either KT or NMS conductivities are used in the calculation. This value of $\sigma(E_1)$ is too large for hopping by orders of magnitude.

The problem now is to explain these transport results in terms of a model for the molecular and electronic structure. Nakamura, Matsumura, and Shimoji have suggested for Tl_xS_{1-x} that there is a mixture of Tl^+ and Tl^{+3} ions for $x < \frac{2}{3}$, and transport occurs by charge exchange between these ions. One would expect appreciable dependence of σ on x in such a mechanism, whereas σ is independent of x. In addition, it seems likely that the time scale for charge exchange would be the vibrational period as in hopping, whereas the magnitude of $\sigma(E_1)$ indicates band transport. Although an ionic model seems reasonable for Tl_2S, we think it is likely that the excess sulfur is largely covalently bonded to itself to form chain molecule ions of the form $(S_n)^{2-}$ rather than in the form of ionic Tl_2S_3. (There is no compound Tl_2S_3 in the Tl–S phase diagram, although Tl_2Se_3 and Tl_2Te_3 both occur.) Assuming that excess S or Se is covalently bonded in chains, as we have suggested for Te, the same type of bond equilibrium should occur as discussed in Section 7.3, and the concentration of dangling bonds N_d should increase with decreasing x. Since the electronic screening is weak, the dangling bonds would form acceptor states above the valence band edge which are largely unoccupied (by holes). In the case of Se–Te alloys, there is evidence that the hole concentration increases rapidly with N_d, and this has been explained by a rapid decrease in the energy E_a of the acceptor states. Since $E_f - E_{v1} \sim (E_a - E_{v1})/2$, the result is a rapid increase in σ as N_d increases. In the present case, σ remains nearly constant in spite of an expected large increase in N_d, which needs to be explained. The value of $E_a - E_{v1}$ is determined by several factors: the distance between the acceptor states, the density of mobile screening charge, and the high frequency dielectric constant K. The situation is similar for the first two factors, but quite different for the third. In the case of Se_xTe_{1-x}, a decrease in x not only increases N_d but also increases the concentration of tellurium. Te is relatively large and polarizable, so that K increases. In the present case, a decrease in x is accompanied by a reduction in K due to the large, polarizable Tl atoms. We think that the relatively small change in σ with x may be explained by the opposing effects of decreasing K and increasing N_d. The proposed model may also be able to account for the discrepancies between E_σ and E_{S0} in in terms of an increase in the conductivity at the mobility edge with T as a result of overlap between the acceptor band and the valence band, as in the case of Se–Te.

Petit and Camp (1975) have reported an interesting study of $\sigma(T)$ for various x in Tl_xSe_{1-x} in the range $x \leq 0.38$. They found that the behavior is described by $\sigma = \sigma_0 \exp(-E_A/kT)$, where E_A and σ_0 vary systematically with x. The quantities E_A and σ_0 at $x = 0.002$ differ greatly from the values for pure Se, which shows that the addition of small amounts of Tl introduces a different transport mechanism. As x increases from 0.002 to 0.38, E_A decreases moderately from 0.53 eV to 0.43 eV, but σ_0 increases rapidly from 0.44 to 17,000 ohm^{-1} cm^{-1}. The lower values of σ_0 correspond to hopping, and the relatively large value of E_A indicates that E_f is above the energy of the hopping states. Petit and Camp make an analysis of the dependence of σ_0 on x and show that in the limit of small x, $\ln \sigma_0$ varies linearly with the average distance r_s between the Tl atoms. This corresponds to Eq. 6.18 in which R is replaced by r_s. The authors use a more sophisticated version of this equation which is based on percolation theory (Ambegaokar et al., 1971). It has the effect of replacing $2\alpha R$ by $1.4\alpha r_s$, and they determine from the slope that the attenuation distance $\alpha^{-1} = 2.5$ Å. Deviations from the linear relation between $\ln \sigma_0$ and r_s occur when $r_s < 6$ Å which corresponds to $\sigma_0 > 55$ ohm^{-1} cm^{-1} and $x \gtrsim 0.02$.

It is interesting to speculate on the mechanism which leads to hopping sites at the Tl atoms. Since the Tl–Se bonds are ionic or strongly polar, the Se atoms next to the Tl atoms can be expected to have some negative charge, which would cause the valence band edge to shift upward. At low Tl concentrations, these potential fluctuations could give rise to localized states above the valence band edge which trap holes. We suggest that these states are responsible for the hopping transport. At larger x, the trapping states becomes shallower and there is a gradual shift to transport in extended states, corresponding to the larger values of σ_0.

8.8 As_2Se_2 AND RELATED ALLOYS

These alloys have particular importance because they form natural glasses, and their properties can be followed to low temperatures—below the glass transition temperature T_g—without a sharp change in the molecular structure. As illustrated by the behavior of σ and μ_H of several of these alloys, shown in Fig. 2.13, the electronic properties seem to change continuously between the liquid and vitreous states. Edmond (1966) has made an extensive study of As_2Te_3–As_2Se_3 alloys in the vitreous and liquid states, including measurements of σ, S, and optical absorption. Similar measurements on the glassy and liquid phases of As_xSe_{1-x} have been reported in the range $0.3 < x < 0.5$ by Hurst and Davis (1974a, b). A study of S and σ in liquid As_xTe_{1-x} for $x < 0.55$ has been made by Oberafo (1975). Both the As–Se and As–Te alloys are distinguished by the absence of a singularity in the behavior of σ and S at the stoichiometric composition. We take this as evidence that excess arsenic is bonded covalently as well as excess chalcogenide, as discussed in Section 8.1.

The function $\sigma(x)$ has a maximum at $x = 0.40$ in vitreous As_xSe_{1-x} rather than a minimum, and Hurst and Davis (1974a) have traced the cause to a minimum in the mobility gap. Since the bond orbital splitting for As–Se is expected to be larger than for As–As or Se–Se bonds (see Section 8.1), one expects the band gap to be a maximum at As_2Se_3. Hurst and Davis point out that the greater compositional disorder at other compositions should increase the distance of the mobility edge from the band edge, and this can account for the minimum in the mobility gap at As_2Se_3.

Plots of S and $\ln \sigma$ versus $1/T$ in As_xSe_{1-x} are linear in both the liquid and glass. Experimental difficulties prevent measurements through the glass transition temperature, but extrapolations of the curves from either side seem to meet at T_g. The activation energy E_σ for the conductivity is larger in the liquid, and this is explained by an increase in $-dE_G/dT$ at T_g. This explanation is suggested by the behavior of the band gap E_G as inferred from optical absorption measurements made continuously through T_g. The activation energy E_S of the thermopower in the glass is only slightly different from E_σ, and the average of E_S and E_σ is shown as a function of composition in Fig. 8.32.

In terms of Eqs. 6.5, the S and σ data for the glasses yield $\sigma(E_1) \sim 10$ ohm^{-1} cm^{-1}, which is an order of magnitude less than the value expected

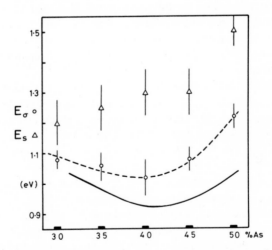

Fig. 8.32. Compositional dependence of the activation energies for the conductivity E_σ and the thermopower E_S in As–Se alloys. The experimental points refer to the liquid, and the solid line describes the average of E_σ and E_S for the glass. (Hurst and Davis, 1974b).

at the mobility edge. Hurst and Davis suggest that transport occurs at the mobility edge for both electrons and holes, with the latter dominating. A relatively small contribution by electrons reduces S from S_p, which accounts for the apparently small value of $\sigma(E_1)$. This also provides an explanation for the fact that E_S is somewhat larger (~ 0.1 eV) than E_σ. In this model, then, E_f is pinned near the center of the band gap by the equilibrium between the thermally excited electron-hole pairs, and a slight asymmetry in the distance of the mobility edges can account for the p-type behavior. The model suggests that $E_\sigma = E_f - E_{v1}$ (at $T \to 0$) should be equal to one-half the optical band gap, and this relationship is obeyed rather well at all compositions. Thus the model seems to provide a good explanation for the behavior of the glass.

The behavior of the electrical conductivity of liquid As_xSe_{1-x} (Hurst, 1974b) seems to be a reasonable extension of the behavior of the glass. The quantity E_σ is 0.1–0.2 eV larger than in the glass, as shown in Fig. 8.32. But E_S, shown also in the figure, is a good deal larger than E_σ, and the reason for this discrepancy is not clear. In addition, the plots of S versus $1/T$ show a decrease in slope at high T in the Se-rich alloys. Hurst and Davis suggest that this change in slope may be related to structural changes in the liquid, as indicated by a change in slope of the $\ln \eta$ (viscosity) versus $1/T$ and the disappearance of a vibrational optical absorption band in liquid As_2Se_3 (Taylor *et al.*, 1971), mentioned in Section 6.5, which occurs in the same temperature range. It is not clear why structural changes should affect S and not σ, but one possibility, suggested by Hurst and Davis, is a heat of transport H which would add a term eH/T to S (Fritzsche, 1971). This may be sensitive to the molecular structure, and it can also explain the difference between E_S and E_σ. Hurst and Davis suggest ambipolar transport as another possible explanation for the anomalous behavior of the thermopower.

The thermoelectric behavior of As_xTe_{1-x} reported by Oberafo (1975) is similar to that of As_xSe_{1-x} in showing a continuous change as a function of x. In this case, since Te is near-metallic, the change is monotonic. Little analysis has been made of these results. Oberafo observes nearly the same activation energies for the thermopower ($E_S = 0.73$ eV) and the electrical conductivity ($E_\sigma = 1.02$ eV) for As_2Te_3 as reported previously by Edmond (1966). We note that in contrast to the data of Hurst and Davis for liquid As_xSe_{1-x}, Oberafo's measurements are in a range of σ (10–2000 ohm^{-1} cm^{-1}) where we expect an acceptor band due to dangling bonds to merge with the valence band as T and σ increases, as discussed in Section 8.6.2. This would account for the large discrepancy between E_σ and E_S. Edmond's plots of $\ln \sigma$ versus $1/T$ for As_2Se_3, As_2Se_2Te, and As_2SeTe_2 all show an increase in the activation energy in the region of σ between 0.1 to 10 ohm^{-1}

cm^{-1} and an overall behavior which is similar to the curves for Se$_x$Te$_{1-x}$ shown in Fig. 2.5, which again seems to be a manifestation of the same phenomenon.

The behavior of Sb–Se alloys is different from arsenic chalcogenides in that there is a p–n transition at Sb$_2$Se$_3$ between alloys containing excess Se and excess Sb (Kazandzhan, 1968). The thermopower of Sb$_2$Se$_3$ is nearly zero over a wide temperature range, and σ is at a minimum at that composition. Since S increases and σ decreases relatively slowly with added Sb, and both parameters change strongly with T when there is excess Sb, we suppose that the excess Sb is still largely covalently bonded. But the n-type behavior suggests that the excess Sb is converted to a metallic form (i.e., ions plus electrons) when the bonds are broken. This is in contrast to the behavior of As alloys, where broken As–As bonds apparently yield dangling bonds and p-type behavior.

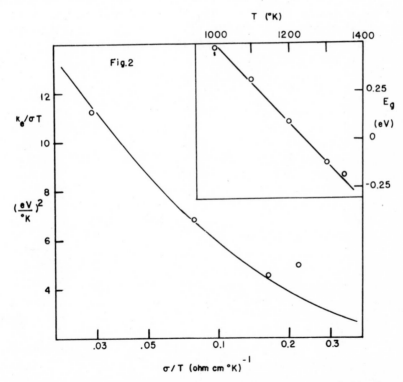

Fig. 8.33.   Analysis of Sb$_2$Se$_3$ data in terms of the parabolic symmetrical band model. The main graph shows a fit of the theoretical curve for $\kappa_e/\sigma T$ versus σ/T to the experimental points with $\kappa_e = \kappa - 1.0 \times 10^{-3}$ cal/cm deg. The insert shows the derived dependence of the band gap on temperature (Cutler, 1974b).

The thermal conductivity of Sb_2Se_3 has been measured by Fedorov and Machuev (1968). This is another case where the excess electronic κ_e has been analyzed incorrectly by using the Maxwell–Boltzmann approximation in the expression for the ambipolar contribution as discussed in Section 6.3. This has led to speculation about other mechanisms for the thermal transport (Shadrichev and Smirnov, 1971; Regel et al., 1972). The fact that $S(T)$ is nearly zero suggests that the pseudogap is nearly symmetrical, and the behavior of κ_e has been accounted for in terms of the model of symmetric parabolic bands, discussed in Section 6.3 (Cutler, 1974b). In addition to $\kappa_e/\sigma T$ versus E_G/kT as shown in Fig. 6.2, the model yields σ/T to within the unknown constant B in $\sigma(E) = BE$, so that a comparison between theory and experiment is possible by plotting $\ln(\sigma/T)$ versus $\kappa_e/\sigma T$. This is shown in Fig. 8.33, and it is seen that the experimental points agree well with the theoretical curve with $B = 900$ ohm^{-1} cm^{-1}. The quantity B is ~ 3 times smaller than the values obtained for Tl–Te alloys. Since $E_f = -E_G/2$, a knowledge of B permits the determination of the band gap E_G versus T, which is shown in the inset of Fig. 8.33. The value of $dE_G/dT = -2.0 \times 10^{-3}$ eV/deg is comparable to that obtained by Edmond (1966) for As_2Se_3 (-1.7×10^{-3} eV/deg).

PHENOMENOLOGICAL
EXPRESSIONS
FOR σ, S, AND κₑ

If the electrical conductivity σ is the sum of contributions at different energies $\sigma(E)$ according to the formula

$$\sigma = -\int_0^\infty \sigma(E)(\partial f/\partial E)\, dE, \tag{A1}$$

and if the heat of transport of a charge at energy E is $E - E_f$, then the thermopower S and the electronic thermal conductivity κ_e can be expressed in terms of moments of $(E - E_f)/kT$ defined by

$$\left\langle \left(\frac{E - E_f}{kT}\right)^n \right\rangle = -\int_0^\infty \frac{\sigma(E)}{\sigma}\left(\frac{E - E_f}{kT}\right)^n \frac{\partial f}{\partial E}\, dE. \tag{A2}$$

This conclusion is arrived at by use of the Kelvin equation $\Pi = ST$, where Π is the Peltier coefficient (Cutler, 1972; Fritzsche, 1971; Mott and Davis, 1971).

Since the isothermal current density $J = \sigma F$, where F is the electric field, the differential current density at energy E is

$$dJ = -F\sigma(E)(\partial f/\partial E)\, dE. \tag{A3}$$

The differential heat flux is then

$$dQ = -(1/e)(E - E_f)\, dJ = (F/e)(E - E_f)\sigma(E)(\partial f/\partial E)\, dE. \tag{A4}$$

Since $Q = \Pi J$ at constant T, integration of Eq. A4 together with the Kelvin relation for Π gives

$$S = -\left(\frac{k}{e}\right)\left\langle \left(\frac{E - E_f}{kT}\right)\right\rangle. \tag{A5}$$

Consider now J and Q in the presence of a temperature gradient ∇T as well as an electrical field, making use of Eq. A5 to express S. The phenomenological equations are

$$J = \sigma F - S\sigma \, \nabla T,$$
$$Q = \Pi \sigma F - \kappa_e{}^* \, \nabla T. \tag{A6}$$

J can be expressed in terms of $\sigma(E)$ with the use of Eqs. A1 and A5. Then, following the same procedure as in Eqs. A3 and A4 and integrating dQ, one arrives at a result for Q which shows on comparison with Eq. A6 that

$$\kappa_e{}^* = \frac{k^2}{e^2} \left\langle \left(\frac{E - E_f}{kT} \right)^2 \right\rangle \sigma T. \tag{A7}$$

Since $\kappa_e = Q/\nabla T$ when $J = 0$, Eqs. A6 leads to the result $\kappa = \kappa^* - S\Pi\sigma$, so one can write

$$\kappa_e = \left(\frac{k}{e} \right)^2 \left[\left\langle \left(\frac{E - E_f}{kT} \right)^2 \right\rangle - \left\langle \left(\frac{E - E_f}{kT} \right) \right\rangle^2 \right] \sigma T. \tag{A8}$$

FERMI–DIRAC INTEGRALS AND TRANSPORT CALCULATIONS

B1 FERMI–DIRAC INTEGRALS

In the intermediate range of values of E_f/kT, where neither the Maxwell–Boltzmann (MB) nor the metallic approximation is valid, Fermi–Dirac integrals or similar functions must be used to evaluate the electron (or hole) density and transport parameters. Most existing data for liquid semiconductors lie in this intermediate range, and application of the MB or metallic approximation often leads to misleading or excessively crude results.

Fermi–Dirac integrals have been discussed in several places, and a number of tabulations of their values have been prepared (Blakemore, 1962). It is worth while to discuss here some problems incidental to their use in analyzing data for liquid semiconductors. The advent of computer techniques makes it practical to use more complicated related definite integrals which depend on more than one parameter, and we discuss some of these ramifications.

The Fermi–Dirac integral $F_n(\xi)$ is defined by

$$F_n(\xi) = \int_0^\infty \frac{x^n \, d\xi}{e^{x-\xi} + 1}. \tag{B1}$$

These integrals arise when equations such as A1 and A5 are applied to problems in which $\sigma(E)$ is proportional to a power of E. If $\sigma(E) = BE^n$, the change in variable $E = xkT$ and $E_f = \xi kT$ yields

$$\sigma = -B(kT)^n \int_0^\infty x^n (\partial f/\partial x) \, dx, \tag{B2}$$

where $f = [1 + \exp(x - \xi)]^{-1}$. Integration by parts yields

$$\sigma = B(kT)^n n F_{n-1}(\xi). \tag{B3}$$

In a similar manner, Eq. A5 leads to

$$S = -\left(\frac{k}{e}\right)\left[\frac{(n + 1)F_n(\xi)}{nF_{n-1}(\xi)} - \xi\right]. \tag{B4}$$

If the density of states is given by $N(E) = CE^m$ [$m = n/2$ if the diffusive model for $\sigma(E)$ is valid], ξ is related to the electron density n by

$$n = \int_0^\infty fCE^m \, dE = C(kT)^{m+1} F_m(\xi).$$ (B5)

Tables of $F_n(\xi)$ have been published for various integral and half-integral values of n in the range $-4 \leq \xi \leq 10$ (Blakemore, 1962). Beyond these limits for ξ, the MB or the metallic approximation is good:

$$F_n(\xi) \cong n! e^\xi \qquad \text{if} \quad \xi < -4,$$ (B6a)

$$F_n(\xi) \cong \xi^{n+1}/(n+1) \quad \text{if} \quad \xi > 10.$$ (B6b)

If a computer is used in making calculations, it is much simpler to calculate Fermi–Dirac integrals as needed, rather than to refer to tables. In Section B4 a set of subroutines are listed which the author has found useful for this purpose. The accuracy seems to be generally better than 0.01%. But if the integrals are used in formulas requiring subtraction of nearly equal quantities, as, for example, in Eq. B4 for $\xi > 1$, the ultimate error may be much larger.

B2 CURVE FITTING

The classic way to fit experimental points to a theoretical curve is to express the experimental information in terms of functions which can be plotted to yield a straight line, and the unknown constants of the theory are deduced from the slope and intercept. This is not possible for expressions containing nonanalytical functions, and the next best method is to express the theoretical relation in the form

$$E_2 = G_1(C_1 E_1)$$ (B7a)

or

$$G_2(C_2 E_2) = G_1(C_1 E_1),$$ (B7b)

where C_1 and C_2 are unknown constants, E_1 and E_2 are numbers which can be determined experimentally, and G_1 and G_2 are the nonanalytical functions which can be calculated. In the first case, the theoretical relation is compared with experiment using a semilog plot, so that the log scale represents $C_2 E_2$ for the theoretical curve and E_2 for the experimental curve. If the two curves can be superposed by shifting along the logarithmic axis, the value of C_2 is derived. In the case of Eq. B7b, both scales are logarithmic, and shifts are made in both directions to determine C_1 and C_2. If the curve is a straight line, of course, one cannot infer unique values for C_1 and C_2.

The theoretical curves in Fig. 7.1 are examples of a test of a relation of the type B7a. It was obtained by combining Eqs. B3 and B4 so that $C_1 E_1 = \sigma/B(kT)^n$, $E_2 = -S$, and G_1 is the function obtained from the two equations when the implicit variable ξ is eliminated. The theoretical curves correspond to $n = \frac{3}{2}$, 1, and $\frac{1}{2}$, and each was shifted along the log σ axis to superpose them as well as possible on the experimental plot of log σ versus S.

B3 MOBILITY EDGE MODEL

It is a simple matter to extend the equations of Section B1 to the case where $\sigma(E)$ is assumed to drop to zero at an arbitrary energy E_c. This causes the lower limit of the integrals in Eq. A2 to be E_c instead of zero, and when the variable of integration is $x = E/kT$, the lower limit is $x_c = E_c/kT$. Therefore, a modified Fermi–Dirac integral is defined as

$$G_n(\xi, x_c) = -\int_{x_c}^{\infty} x^n(\partial f/\partial x)\, dx. \tag{B8}$$

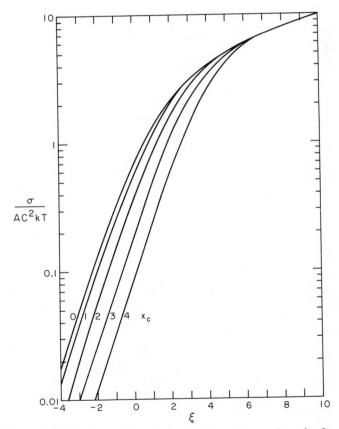

Fig. B1. Theoretical curves for σ versus ξ at various constant values of x_c for a mobility edge model in which $\sigma(E) \propto E$.

It is easy to see that $G_n(\xi, 0) = nF_{n-1}(\xi)$, so that the theoretical relations for σ, S, etc., are generated by replacing $F_n(\xi)$ by $G_{n+1}(\xi, x_c)/(n+1)$.

It has been useful to prepare families of curves which show the relations of σ, S, ξ, and n for various values of x_c. One must of course specify $\sigma(E)$. For the case where $N(E) = CE^{1/2}$ and $\sigma(E) = AC^2E$, the result is

$$\sigma = AC^2kTG_1(\xi, x_c), \tag{B9a}$$

and

$$S = -\left(\frac{k}{e}\right)\left[\frac{G_2(\xi, x_c)}{G_1(\xi, x_c)} - \xi\right]. \tag{B9b}$$

The mobility edge does not affect the equation for n, so that ξ is related to the electron density by Eq. B5.

Plots of $\ln[\sigma/AC^2kT]$ and S versus ξ are shown for various values of x_c in Figs. B1 and B2. They show how σ and S deviate from the MB approximation in Eq. 6.5 (where $(E_1 - E_f)/kT$ is the same as $x_c - \xi$) when $x_c - \xi$

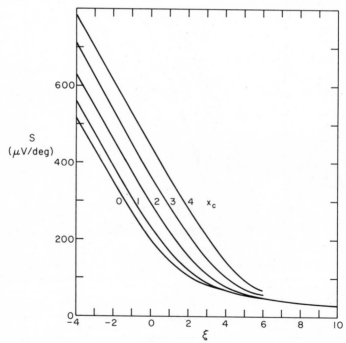

Fig. B2.   Theoretical curves for S versus ξ at various constant values of x_c for a mobility edge model in which $\sigma(E) \propto E$.

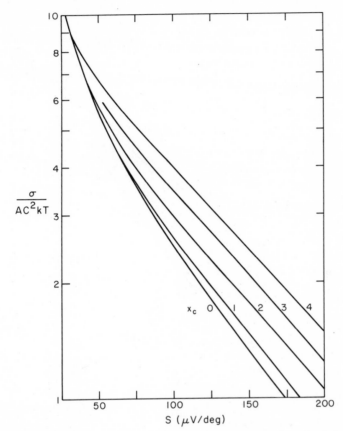

Fig. B3. Theoretical curves for σ versus S at various constant values of x_c for a mobility edge model in which $\sigma(E) \propto E$.

becomes small. In Fig. B3, $\ln[\sigma/AC^2kT]$ is plotted versus S for various values of x_c.

If the electron density n can be be inferred from experimental information, it is useful to examine the dependence of σ or S on n. Plots of $\ln(\sigma/AC^2kT)$ and S versus $\log[n/C(kT)^{3/2}]$ are shown in Figs. B4 and B5. In contrast to Fig. B3, differing values of x_c yield different slopes, so that a study of S or σ versus n could more readily lead to a definitive evaluation of E_c.

B4 COMPUTER SUBROUTINES

These subroutines are written in FORTRAN IV. To calculate $F_n(\xi)$, use the statement CALL FDINT (ANU, XI, F), where n is ANU, ξ is XI, and $F_n(\xi)$ is F. To

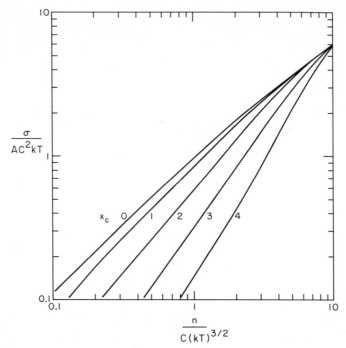

Fig. B4.   Theoretical curves for σ versus n at various constant values of x_c for a mobility edge model in which $\sigma(E) \propto E$.

calculate $G_n(\xi, x_c)$, use the statement CALL MSINT (AN, XI, XC, G), where n is AN, ξ is XI, x_c is XC, and $G_n(\xi, x_c)$ is G.

```
SUBROUTINE GMS(XI,XC,F,A,G)
ETA=XI-XC
XMAX=ETA+10.
IF(XMAX.LT.10.)3,4
3 XMAX=10
4 H1=.05
H2=(XMAX-2.)/100.
X=XC+.05
XA=XC+.025
PHA=1./(EXP(XA-XI)+1.)
PHB=1./(EXP(X-XI)+1.)
F1=F(XA,A)*PHA*(1.-PHA)+.5*F(X,A)*PHB*(1.-PHB)
DO 1 L=2,40
X=X+H1
PHI=1./(EXP(X-XI)+1.)
TERM=F(X,A)*PHI*(1.-PHI)
1 F1=F1+TERM
F1=F1-TERM/2.
F2=TERM/2.
DO 2 M=1,100
X=X+H2
PHI=1./(EXP(X-XI)+1.)
TERM=F(X,A)*PHI*(1.-PHI)
```

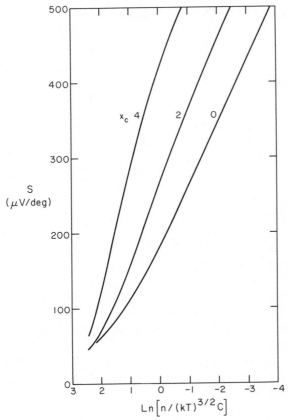

Fig. B5.   Theoretical curves of S versus n at various constant values of x_c for a mobility edge model in which $\sigma(E) \propto E$.

```
2 F2=F2+TERM
  F2=F2-TERM/2.
  G=F1*H1+F2*H2
  RETURN
  END
  SUBROUTINE FDINT(ANU,XI,F)
  CALL MSINT(ANU+1.,XI,0.,G)
  F=G/(ANU+1.)
  RETURN
  END
  SUBROUTINE MSINT(AN,XI,XC,G)
  EXTERNAL PX
  CALL GMS(XI,XC,PX,AN,G)
  RETURN
  END
  FUNCTION PX(X,AN)
  PX=X**AN
  END
```

BOND EQUILIBRIUM THEORY

The independent bond model for molecular equilibrium, according to which the physical properties of a given bond are independent of the other constituents attached to the atoms, is a common approximation in polymer theory. Modifications can be made to distinguish separately larger groupings of atoms whose behavior is independent of configurations outside of that group. Expressions can be derived by application of the law of mass action to either molecules or to bonds, with apparently equivalent results. Equations based on molecular equilibrium have a more direct theoretical justification, but the bond equilibrium approach is simpler and seems to be more versatile. We apply both methods to Tl–Te bonding and show that they are equivalent in that case, and then use the simpler bond equilibrium approach to consider the behavior of Te–Se alloys.

C1 MOLECULAR EQUILIBRIUM THEORY FOR Tl–Te

The basic assumptions of the independent bond model are: (1) each Tl atom is bonded to a Te atom; (2) the equilibrium constant for breaking a Te–Te bond is independent of what other constituents, if any, are attached to the Te atoms making the bond. At equilibrium, the molecular species are described in C1.

Constituent	Structure	Concentration (per atom)	
A_n	$Tl-(Te)_n-Tl$	a_n	
B_n	$Tl-(Te)_n*$	b_n	(C1)
C_n	$*(Te_n)*$	c_n	

where $(Te)_n$ represents a chain of n bonded Te atoms, and the asterisks represent dangling bonds. The equilibrium for the various species is determined by two equilibrium constants:

$$X\text{–Te–Te–}Y \stackrel{k}{=} X\text{Te*} + Y\text{Te*}, \tag{C2}$$

and

$$\tfrac{1}{2}\text{Te}_2 + X\text{–(Te)}_{n-1}\text{–}Y \stackrel{K}{=} X\text{–(Te)}_n\text{–}Y. \tag{C3}$$

The value of k in Eq. C2 is independent of X and Y except for symmetry factors. If $X = Y$, $k = 2k_0$, and if $Y = *$, $k = \tfrac{1}{2}k_0$. It will be seen that Eq. C3 serves only to relate the values of a_n, b_n, and c_n to the chemical potential of Te, so that the Te_2 can be assumed, for convenience, to be a molecule in the gas phase, and the concentration (activity) of the Te_2 will be denoted by t^2.

Applying Eq. C3 to any of the three species A_n, B_n, or C_n gives

$$a_{n+1}/a_n = b_{n+1}/b_n = c_{n+1}/c_n = Kt, \tag{C4}$$

so that by iteration

$$a_n = (Kt)^{n-1}a_1, \qquad b_n = (Kt)^{n-1}b_1, \qquad c_n = (Kt)^{n-1}c_1. \tag{C5}$$

Applying reaction C2 to dissociation of the species A_2 gives

$$b_1{}^2 = 2k_0 a_2, \tag{C6}$$

and applying the reaction C2 to form $C_n + B_1$ from B_{n+1} gives

$$c_n = k_0 b_{n+1}/2b_1. \tag{C7}$$

On combining Eqs. C5, C6, and C7, the result is

$$a_n = y^{n-1}L, \qquad b_n = y^{n-1}(2k_0 Ly)^{1/2}, \qquad c_n = y^n k_0/2, \tag{C8}$$

where we use the symbol y for Kt and L for a_1, the concentration of Tl_2Te. It is seen that K and t always appear together, which reflects the arbitrary role of Eq. C3 in describing the chemical potential of Te. The chemical potential of Tl is reflected in the value of L. The values of L and y are thus governed by the constraints

$$x = \sum_{n=1}^{\infty} (2a_n + b_n), \tag{C9}$$

$$1 - x = \sum_{n=1}^{\infty} n(a_n + b_n + c_n), \tag{C10}$$

respectively. The dangling bond concentration c is given by

$$c = \sum_{n=1}^{\infty} (b_n + 2c_n). \tag{C11}$$

In the original work (Cutler, 1971b) the results were complicated by exclusion of c_1 from the sums. This was done because the independent bond assumption is probably a poor approximation when the species *Te* is formed. However, the experimental range is one where the above equations yield very small values of a_1 so that this restraint unnecessarily complicates the final mathematical result.

Substituting Eqs. C8 into Eqs. C9, C10, and C11 yields infinite sums which contain series expansions of $(1 - y)^{-1}$, so that the results can be written

$$x = [2L + (2k_0Ly)^{1/2}]/(1 - y) \tag{C12}$$

$$1 - x = [L^{1/2} + (k_0y/2)^{1/2}]^2/(1 - y)^2 \tag{C13}$$

$$c = [(2k_0Ly)^{1/2} + k_0y]/(1 - y). \tag{C14}$$

It is not hard to eliminate L and y from these three equations to derive the law of mass action for dangling bonds:

$$\frac{c^2}{1 - 3/2x - c/2} = 2k_0. \tag{C15}$$

Other useful relations give the values of y and L in terms of c and x:

$$y = 1 - \frac{(x + c)}{2(1 - x)}, \tag{C16}$$

$$L = \frac{x^2}{2(1 - x)}. \tag{C17}$$

Equation C17 has the interesting implication that the concentration of Tl_2Te in equilibrium with the other species is independent of temperature, and depends only on composition. Equation C16 has the interesting interpretation that y is the fraction of Te bond ends (whose total concentration is $2 - 2x$) which are tied to each other rather than to Tl atoms or are dangling. The main result, Eq. C15, also has a simple interpretation in terms of equilibrium theory for bonds which will be evident in the discussion of that method in the next section.

We have already indicated how restraints can be introduced into the independent bond equations by eliminating some species. A more general method is to introduce a different equilibrium constant for certain species. As an example, a proximity effect can be introduced by assuming a different free energy for breaking a bond when a Te atom is attached to a Tl atom. The simplest self-consistent model for this leads to a modified equilibrium constant for the reaction

$$B_1 + \tfrac{1}{2} Te_2 \overset{k\beta}{=\!=} B_2, \tag{C18}$$

where $\beta > 1$ implies that the B_1 species is relatively unstable. This has the result that

$$b_1 = (2k_0Ly)^{1/2}/\beta, \tag{C19}$$

whereas b_n is given in Eq. C8 for $n > 1$. The effect on the solution is to add a term $(\beta^{-1} - 1)(2k_0 yL)^{1/2}$ to the right-hand side of each of Eqs. C12, C13, and C14. With this modification, elimination to L and y leads to a much more complicated relation for c than Eq. C15.

C2   BOND EQUILIBRIUM METHOD

In this method, dangling bonds are treated as a chemical entities with concentration c, which can combine to form bonds, with a concentration b. To formulate the law of mass action, it is useful to consider the concentration h of Te–Te half-bonds. For $Tl_x Te_{1-x}$,

$$h = 2(1 - x) - x - c, \tag{C20}$$

so that

$$b = h/2 = 1 - 3x/2 - c/2. \tag{C21}$$

The law of mass action for the reaction

$$\text{bond} \stackrel{\kappa}{=} 2 \text{ dangling bonds} \tag{C22}$$

gives

$$\kappa = \frac{c^2}{(1 - 3x/2 - c/2)}. \tag{C23}$$

Comparison with Eq. C15 shows that $\kappa = 2k_0$.

This formulation has potential value in consideration of equilibrium involving atoms with valence greater than two, where the resulting molecules are more complicated than chains. The molecular equilibrium approach requires that every type of molecular species be identified and accounted for, and the mathematical expressions which result are likely to be intractible. The bond equilibrium method requires a description of all types of half-bond species and their concentrations. When threefold bonding occurs, this may not be a trivial mathematical problem either, but it seems to be much simpler than the results of the molecular equilibrium method.

C3   INDEPENDENT BOND MODEL FOR Se–Te

Let us apply bond equilibrium method to the problem of the alloy $Se_x Te_{1-x}$, ignoring the possibility of ring formation. In this case, there are two dangling bond species, Te with concentration c_T, and Se with concentration c_S. There are three types of bonds with concentrations b_{TT}(Te–Te), b_{TS}(Te–Se), and b_{SS}(Se–Se).

The half-bond concentrations are

$$h_S = 2x - c_S, \qquad h_T = 2(1 - x) - c_T, \tag{C24}$$

and they are related to the bond concentrations by

$$h_S = 2b_{SS} + b_{TS}, \qquad h_T = 2b_{TT} + b_{TS}. \tag{C25}$$

The law of mass action for the three types of bonds gives

$$K_{TT} = c_T{}^2/b_{TT}, \qquad K_{TS} = c_T c_S/b_{TS}, \qquad K_{SS} = c_S{}^2/b_{SS}. \tag{C26}$$

Let us define a new parameter k_m by

$$k_m = K_{TS}/(K_{SS}K_{TT})^{1/2}. \tag{C27}$$

Replacing K_{TS} by k_m, it is a simple matter to derive from Eqs. C24–27 the following relations which have been put in a symmetrical form:

$$K_{TT} = 2c_T{}^2/(2 - 2x - c_T - b_{ST}), \tag{C28}$$

$$K_{SS} = 2c_S{}^2/(2x - c_S - b_{ST}), \tag{C29}$$

$$b_{ST}{}^2 = (2x - b_{ST} - c_S)(2 - 2x - b_{ST} - c_T)/4k_m{}^2. \tag{C30}$$

with

$$2b_{SS} = 2x - c_S - b_{ST}, \tag{C31}$$

$$2b_{TT} = 2 - 2x - c_T - b_{ST}. \tag{C32}$$

Equations C28, C29, and C30 need to be solved to determine c_S and c_T for a given set of values of the equilibrium constants K_{SS}, K_{TT}, and k_m.

Relatively simple results can be obtained in the limit where c_S and c_T are relatively small. Equation C30 can then be written

$$b_{ST}{}^2(4k_m{}^2 - 1) + 2b_{ST} - 4x(1 - x) = 0. \tag{C33}$$

There are evidently two interesting limits for k_m; $k_m = \frac{1}{2}$ corresponds to purely random bonding and $k_m \ll 1$ corresponds to a preference for Se–Te bonding. For these two situations, b_{ST} has a simple dependence on x:

$$b_{ST} = 2x(1 - x) \qquad\qquad \text{when}\quad k_m = \tfrac{1}{2}, \tag{C34a}$$

$$b_{ST} = \begin{cases} 2 - 2x & \text{if}\quad x > \tfrac{1}{2} \\ 2x & \text{if}\quad x < \tfrac{1}{2} \end{cases} \qquad \text{when}\quad k_m = 0. \tag{C34b}$$

In the same limit (small c_S and c_T), Eq. C28 and C29 yield for c_T and c_S

$$c_T = [1 - x - b_{ST}/2]^{1/2} K_{TT}{}^{1/2}, \tag{C28a}$$

$$c_S = [x - b_{ST}/2]^{1/2} K_{SS}{}^{1/2}, \tag{C29a}$$

where Eq. C34a or C34b can be substituted for b_{ST} for extreme values of k_m, or the solution of Eq. C33 in the intermediate case.

REFERENCES

Abragam, A. (1961). "The Principles of Nuclear Magnetism." Oxford Univ. Press, London and New York.

Abrikosov, N. Kh., Chizhevskaya, S. N., and Kurbatov, V. P. (1970). *Ind. Lab.* **36**, 581.

Adams, P. D., and Leach, J. S. (1967). *Phys. Rev.* **156**, 178.

Allgaier, R. S. (1969). *Phys. Rev.* **185**, 227.

Allgaier, R. S. (1970). *Phys. Rev. B* **2**, 2257.

Ambegaokar, V., Halperin, B. I., and Langer, J. S. (1971). *Phys. Rev. B* **4**, 2612.

Amboise, M. D., Handfield, G., and Bourgon, M. (1968). *Can. J. Phys.* **46**, 3545.

Anderson, P. W. (1958). *Phys. Rev.* **109**, 1492.

Andreev, A. A. (1973). *Proc. Int. Conf. Amorphous Liquid Semicond. 5th, Garmisch-Partenkirchen*, 343.

Andreev, A. A., and Mamadaliev, M. (1972). *J. Non-Cryst. Solids* **8–10**, 287.

Ashcroft, N. W., and Lekner, J. (1966). *Phys. Rev.* **145** 83.

Ashcroft, N. W., and Russakoff, G. (1970). *Phys, Rev. A* **1**, 39.

Baker, E. H. (1968). *J. Chem. Soc.* A1089.

Beer, A. C. (1963). "Galvanomagnetic Effects in Semiconductors." Academic Press, New York.

Beer, S. Z. (ed.) (1972). "Liquid Metals, Chemistry and Physics" Dekker, New York.

Berggren, K. -F. (1974). *Phil. Mag.* **30**, 1.

Berggren, K. -F., Martino, F., and Lindell, G. (1974). *Phys. Rev. B* **9**, 4097.

Bhatia, A. B., and Hargrove, W. H. (1973). *Lett. Nuovo Cimento II* **8**, 1025.

Bhatia, A. B., and Hargrove, W. H. (1974). *Phys. Rev. B* **10**, 3186.

Bhatia, A. B., and Thornton, D. E. (1970). *Phys. Rev. B* **2**, 3004.

Bishop, S. G. (1973). *Proc. Int. Conf. Amorphous Liquid Semicond. 5th, Garmisch-Partenkirchen* 997.

Bitler, W. R., Yang, L., and Derge, G. (1957). *J. Appl. Phys.* **28**, 514.

Blakemore, J. S. (1962). "Semiconductor Statistics." Pergamon, Oxford.

Blakeway, R. (1969). *Phil Mag.* **20**, 965.

Bockris, J. O'M., White, J. L., and Mackenzie, J. D. (1954). "Physicochemical Measurements at High Temperatures." Butterworths, London.

Boolchand, P., and Suranyi, P. (1973). *Phys. Rev. B* **7**, 57.

Bredig, M. A. (1964). "Molten Salt Chemistry" (M. Blander, ed.), p. 367. Wiley (Interscience), New York.

Bridgman, P. W. (1961). "The Thermodynamics of Electrical Phenomena and a Condensed Collection of Thermodynamic Formulas," Chapter 2. Dover, New York.

Brown, D., Moore, D. S., and Seymour, E. F. W. (1971). *Phil. Mag.* **23**, 1249.

Busch, G., and Güntherodt, H. J. (1974). *Solid State Phys.* **24**, 237.

Busch, G., and Tièche, Y. (1963). *Phys. Kondens. Mater.* **1**, 78.

Busch, G., and Yuan, S. (1963). *Phys. Kondens. Mater.* **1**, 37.

Butcher, P. N. (1974). *J. Phys. C* **7**, 879.

Cabane, B., and Friedel, J. (1971). *J. Phys.* **32**, 73.

Cabane, B., and Froidevaux, C. (1969). *Phys. Lett.* **A29**, 512.

Castanet, R., Bros, J. P., and Laffitte M. (1968). *J. Chem. Phys.* (Paris) **65**, 1536.

Cate, R. C., Wright, J. G., and Cusack, N. E. (1970). *Phys. Lett.* **A32**, 467.

Clark, A. H. (1967). *Phys. Rev.* **154**, 750.

Cohen, M. H., and Jortner, J. (1973). *Phys. Rev. Lett.* **30**, 699.

Cohen, M. H., and Sak, J. (1972). *J. Non-Cryst. Solids* **8–10**, 696.

Cohen, M. H., and Thompson, J. C. (1968). *Adv. Phys.* **17**, 857.

Coulson, C. A. (1961). "Valence," 2nd ed. Oxford Univ. Press, London and New York.

Cusack, N. E., and Kendall, P. (1958). *Proc. Phys. Soc.* **72**, 898.

Cusack, N. E., Enderby, J. E., Kendall, P. W., and Tièche, Y. (1965). *J. Sci. Instrum.* **42**, 256.

Cutler, M. (1963). Rep. GA-4420, October 23, General Atomic, San Diego, California.

Cutler, M. (1969). *Phys. Rev.* **181**, 1102.

Cutler, M. (1971a). *Phil. Mag.* **24**, 381.

Cutler, M. (1971b). *Phil. Mag.* **24**, 401.

Cutler, M. (1972). *Phil. Mag.* **25**, 173.

Cutler, M. (1973). *Solid State Commun.* **13**, 1293.

Cutler, M. (1974a). *Phys. Rev. B* **9**, 1762.

Cutler, M. (1974b). *Proc. Int. Conf. Phys. Semicond. 12th, Stuttgart* 1066.

Cutler, M. (1976a). *Phil. Mag.* **33**, 559.

Cutler, M. (1976b). *Phys. Rev. B* **15**, 693 (1977).

Cutler, M. (1976c). *Phys. Rev. B* **14**, 5344.

Cutler, M. (1976d). *Can. J. Chem.* (in press).

Cutler, M. (1976e). *J. Non-Cryst. Solids*, **21**, 137.

Cutler, M., and Field, M. B. (1968). *Phys. Rev.* **169**, 632.

Cutler, M., and Mallon, C. E. (1962). *J. Chem. Phys.* **37**, 2677.

Cutler, M., and Mallon, C. E. (1965). *J. Appl. Phys.* **36**, 201.

Cutler, M., and Mallon, C. E. (1966). *Phys. Rev.* **144**, 642.

Cutler, M., and Mott, N. F. (1969). *Phys. Rev.* **181**, 1336.

Cutler, M., and Petersen, R. L. (1970). *Phil. Mag.* **21**, 1033.

Dahl, M. M. (1969). M. S. Thesis, Oregon State Univ. May 8.

Dancy, E. A. (1965). *Trans. Metall. Soc. AIME* **233**, 270.

Dancy, E. A., and Derge, G. (1963). *Trans. Metall. Soc. AIME* **227**, 1034.

Dancy, E. A., Pastorek, R. L., and Derge, G. J. (1965). *Trans Metall. Soc. AIME* **233**, 1645.

Darken, L. S. (1967). *Trans. Metall. Soc. AIME* **239**, 80.

Dixon, A. J., and Ertl, M. E. (1971). *J. Phys. D* **4**, 83.

Donally, J. M., and Cutler, M. (1968). *Phys. Rev.* **176**, 1003.

Donally, J. M., and Cutler, M. (1972). *J. Phys. Chem. Solids* **33**, 1017.

Dow, J. D., and Redfield, D. (1972). *Phys. Rev. B* **5**, 594.

Edmond, J. T. (1966). *Brit. J. Appl. Phys.* **17**, 979.

Edwards, S. F. (1962). *Proc. Roy. Soc.* (*London*) *Ser. A* **267**, 518.

Egan, J. J. (1954). *Acta Metall.* **7**, 560.

Egelstaff, P. A. (1967). "An Introduction to the Liquid State." Academic Press, New York.

Eggarter, T. P., and Cohen, M. H. (1970). *Phys. Rev. Lett.* **25**, 807.

Eisenberg, A., and Tobolsky, A. V. (1960). *J. Polym. Sci.* **46**, 19.

Enderby, J. E., and Collings, E. W. (1970). *J. Non-Cryst. Solids* **4**, 161.

Enderby, J. E., and Hawker, I. (1972). *J. Non-Cryst. Solids* **8–10**, 687.
Enderby, J. E., and Simmons, C. J. (1969). *Phil Mag.* **20**, 125.
Enderby, J. E., and Walsh, L. (1965). *Phys. Lett.* **14**, 9.
Enderby, J. E., and Walsh, L. (1966). *Phil Mag.* **14**, 991.
Enderby, J. E., North, D. M., and Egelstaff, P. A. (1967). *Adv. Phys.* **16**, 171.
Epstein, A. S., Fritzsche, H., and Lark-Horowitz, K. (1957). *Phys. Rev.* **107**, 412.
Even, U., and Jortner, J. (1973). *Phys. Rev. B* **8**, 2536.
Faber, T. E. (1972). "Introduction to the Theory of Liquid Metals." Cambridge Univ. Press, London and New York.
Fedorov, V. I., and Machuev, V. I. (1968). *Sov. Phys.–Semicond.* **2**, 110.
Fedorov, V. I., and Machuev, V. I. (1970a). *Sov. Phys.–Solid State* **12**, 221.
Fedorov, V. I., and Machuev, V. I. (1970b). *Sov. Phys.–Solid State* **12**, 484.
Field, M. B. (1967). M. A. Thesis, Oregon State Univ.
Finkman, E., de Fonzo, A. P., and Tauc, J. (1973). *Proc. Int. Conf. Amorphous Liquid Semicond., 5th Garmisch-Partenkirchen,* 997.
Finkman, E., de Fonzo, A. P., and Tauc. J. (1974). *Proc. Int. Conf. Phys. Semicond., 12th,* Stuttgart, 1022.
Fisher, I. Z. (1959). *Sov. Phys.–Solid State* **1**, 171.
Flynn, C. P., and Rigert, J. A. (1973). *Phys. Rev. B* **7**, 3656.
Flory, P. (1942). *J. Chem. Phys.* **10**, 51.
Franck, E. U., and Hensel, F. (1966). *Phys. Rev.* **147**, 109.
Friedel, J. (1954). *Adv. Phys.* **3**, 446.
Friedman, L. (1971). *J. Non-Cryst. Solids* **6**, 329.
Fritzsche, H. (1971). *Solid State Commun.* **9**, 1813.
Fukuda, N., Yokokawa, Nakai, S., and Niwa, K. (1972). *Trans. Jpn. Inst. Met.* **13**, 352.
Fukuyama, H., Ebisawa, H., and Wada, J. (1969). *Prog. Theor. Phys.* **42**, 494.
Gardner, J. A., and Cutler, M. (1976). *Phys. Rev. B* **14**, 4488.
Gee, G. (1952). *Trans. Faraday Soc.* **48**, 515.
Gelatt, C. D., and Ehrenreich, H. (1974). *Phys. Rev. B* **10**, 398.
Genzel, E. (1953). *Z. Phys.* **135**, 177.
Glasstone, S., Laidler, K. J., and Eyring, H. (1941). "The Theory of Rate Processes." McGraw-Hill, New York.
Glazov, V. M., Chizhevskaya, S. N., and Glagoleva, N. N., (1969). "Liquid Semiconductors." Plenum Press, New York.
Glazov, V. M., and Situlina, O. V. *Akad Nauk. SSSR Chem.* **167**, 587.
Gobrecht, H., Gawlik, D., and Mahdjuri, F. (1971a). *Phys. Kondens. Mater.* **13**, 156.
Gobrecht, H., Mahdjuri, F., and Gawlik, D. (1971b). *J. Phys. C* **4**, 2247.
Gordy, W., and Thomas, W. J. O. (1956). *J. Chem. Phys.* **24**, 439.
Greenwood, D. A. (1958). *Proc. Phys. Soc.* **71**, 585.
Gubanov, A. I. (1965). "Quantum Theory of Amorphous Conductors." Consultants Bureau, New York.
Guggenheim, E. A. (1952). "Mixtures." Oxford Univ. Press (Clarendon), London and New York.
Haisty, R. W. (1967). *Rev. Sci. Instrum.* **38**, 262.
Haisty, R. W. (1968). *Rev. Sci. Instrum.* **39**, 778.
Haisty, R. W., and Krebs, H. (1969a). *J. Non-Cryst. Solids* **1**, 399.
Haisty, R. W., and Krebs, H. (1969b). *J. Non-Cryst. Solids* **1**, 427.
Halder, N. C., and Wagner, C. N. J. (1967). *Phys. Lett. A* **24**, 345.
Halperin, B. I., and Lax, M. (1966). *Phys. Rev.* **148**, 722.
Halperin, B. I., and Lax, M. (1967). *Phys. Rev.* **153**, 802.

Hansen, M., and Anderkov, K. (1958). "Constitution of Binary Alloys." 2nd ed. McGraw-Hill, New York; Supplement I, R. P. Elliot (1965); Supplement II, F. A. Shank (1969).

Harvey, W. W. (1961). *Phys. Rev.* **123**, 1666.

Hawker, I., Howe, R. A., and Enderby, J. E. (1973). *Proc. Int. Conf. Amorphous Liquid Semicond., 5th, Garmisch-Partenkirchen* 85.

Heine, V. (1970). *Solid State Phys.* **24**, 1.

Heine, V., and Weaire, D. (1970). *Solid State Phys.* **24**, 249.

Hensel, F., and Franck, E. U. (1968). *Rev. Mod. Phys.* **40**, 697.

Herman, F., and Skillman, S. (1963). "Atomic Structure Calculations". Prentice-Hall, Englewood Cliffs, New Jersey.

Hildebrand, J. H., and Scott, R. L. (1950). "The Solubility of Non-Electrolytes," 3rd ed. Van Nostrand–Reinhold, Princeton, New Jersey.

Hindley, N. K. (1970). *J. Non-Cryst. Solids* **5**, 17.

Hodgkinson, R. J. (1971). *Phil. Mag.* **23**, 673.

Hodgkinson, R. J. (1973). *Proc. Int. Conf. Amorphous Liquid Semicond., 5th Garmisch-Partenkirchen* 841.

Hodgson, J. N. (1963). *Phil. Mag.* **89**, 735.

Hoshino, H., Schmutzler, R. W., and Hensel, F. (1975). *Phys. Lett.* **A51**, 7.

Hoshino, H., Schmutzler, R. W., Warren, Jr., W. W. and Hensel, F. (1976). *Phil. Mag.* **33**, 255.

Hume-Rothery, W., and Raynor, G. V. (1962). "The Structure of Metals and Alloys." Institute of Metals, London.

Hurst, C. H. and Davis, E. A. (1974a). *J. Non-Crystalline Solids* **16**, 343.

Hurst, C. H., and Davis, E. A. (1974b). *J. Non-Crystalline Solids* **16**, 355.

Ichikawa, K., and Thompson, J. C. (1973). *J. Chem. Phys.* **59**, 1680.

Ilschner, B. R., and Wagner, C. (1958). *Acta Metall.* **6**, 712.

Ioannides, P., Nguyen, V. T., and Enderby, J. E. (1973). *Proc. Int. Conf. Properties Liquid Met., 2nd, Tokyo* 391.

Ioffe, A. F. (1957). "Semiconductor Thermoelements and Thermoelectric Cooling." Infosearch, London.

Ioffe, A. F., and Regel, A. R. (1960). *Progr. Semicond.* **4**, 238.

Joannopoulis, J. D., and Cohen, M. L. (1973). *Phys. Rev. B* **7**, 2644.

Kazandzhan, B. I. (1968). *Sov. Phys.–Semicond.* **2**, 329.

Kazandzhan, B. I., and Selin, Y. I. (1974). *Dokl. Akad. Nauk. SSSR* **216**, 67.

Kazandzhan, B. I., and Tsurikov, A. A. (1973). *Sov. Phys.–Dokl. Phys. Chem.* **210**, 430.

Kazandzhan, B. I., and Tsurikov, A. A. (1974a). *Russ. J. Phys. Chem.* **48**, 429.

Kazandzhan, B. I., and Tsurikov, A. A. (1974b). *Russ. J. Phys. Chem.* **48**, 738.

Kazandzhan, B. I., and Tsurikov, A. A. (1974c). *Akad. Nauk SSSR* **48**, 746.

Kazandzhan, B. I., Lobanov, A. A., Selin, Y. I., and Tsurikov, A. A. (1971). *Sov. Phys.–Dokl. Phys. Chem.* **196**, 4.

Kazandzhan, B. I., Lobanov, A. A., Selin, Y. I., and Tsurikov, A. A (1972). *Sov. Phys.–Semicond.* **5**, 1419.

Kazandzhan, B. I., Razumeichenko, L. A., and Tsurikov, A. A. (1974). *Sov. Phys.–Semicond.* **8**, 219.

Keller, J., and Fritz, J. (1973). *Proc. Int. Conf. Amorphous Liquid Semicond., 5th, Garmisch-Partenkirchen* 975.

Keller, J., and Ziman, J. M. (1972). *J. Non-Cryst. Solids* **8–10**, 111.

Kerlin, A. L., and Clark, W. G. (1975). *Phys. Rev. B* **12**, 3533.

Koningsberger, D. C., van Wolput, J. H. M. C., and Reiter, P. C. U. (1971). *Chem. Phys. Lett.* **8**, 145.

Kubaschewski, O., and Evans, E. L. (1958). "Metallurgical Thermochemistry," 3rd ed. Pergamon, Oxford.

Landau, L. D., and Lifshitz, E. M. (1958). "Statistical Physics" (Transl. by E. Peierls and R. F. Peierls) Pergamon, Oxford.

Lee, D. N. (1971). Ph.D. Thesis, Vanderbilt Univ.

Lewis, G. N., and Randall, M. (1961). "Thermodynamics," 2nd ed. (Revised by K. S. Pitzer and L. Brewer) McGraw-Hill, New York.

Liu, C. C., and Angus, J. C. (1969). *J. Electrochem. Soc.* **116**, 1054.

Lizell, B. (1952). *J. Chem. Phys.* **20**, 672.

Lucas, L. D., and Urbain, G. (1962). *C. R. Acad. Sci. Paris* **255**, 3406.

Luttinger, J. M. (1964). *Phys. Rev.* **135**, A1505.

Maekawa, T., Yokokawa, T., and Niwa, K. (1971a). *J. Chem. Thermodynam.* **3**, 143.

Maekawa, T., Yokokawa, T., and Niwa, K. (1971b). *J. Chem. Thermodynam.* **3**, 707.

Maekawa, T., Yokokawa, T. and Niwa, K. (1972a). *J. Chem. Thermodynam.* **4**, 153.

Maekawa, T., Yokokawa, T., and Niwa, K. (1972b). *J. Chem. Thermodynam.* **4**, 873.

Mahdjuri, F. (1973). *Proc. Int. Conf. Amorphous Liquid Semicond., 5th, Garmisch-Partenkirchen* 1295.

Mahdjuri, F. (1975). *J. Phys.* **C8**, 2248.

Male, J. C. (1967). *Brit. J. Appl. Phys.* **18**, 1543.

Mallon, C. E., and Cutler, M. (1965). *Phil. Mag.* **11**, 667.

Massen, C. H., Weijts, A. G. L. M., and Poulis, J. A (1964). *Trans. Faraday Soc.* **60**, 317.

Miller, A., and Abrahams, E. (1960). *Phys. Rev.* **120**, 745.

Mitchell, D. L., Taylor, P. C., and Bishop, S. G. (1971). *Solid State Commun.* **9**, 1833.

Mooser, E., and Pearson, W. B. (1960). *Progr. Semicond.* **5**, 105.

Morris, V. J. (1971). *Phil. Mag.* **24**, 1221.

Mott, N. F. (1968). *Phil. Mag.* **17**, 1269.

Mott, N. F. (1969). *Phil. Mag.* **19**, 835.

Mott, N. F. (1970). *Phil. Mag.* **22**, 1.

Mott, N. F. (1971). *Phil. Mag.* **24**, 1.

Mott, N. F. (1972a). *Phil. Mag.* **26**, 1015.

Mott, N. F. (1972b). *Adv. Phys.* **21**, 785.

Mott, N. F. (1974a). *Phil. Mag.* **29**, 613.

Mott, N. F. (1974b). "Metal-Insulator Transitions." Taylor-Francis, London.

Mott, N. F., and Davis, E. A. (1971). "Electronic Processes in Non-Crystalline Materials." Oxford Univ. Press (Clarendon), London and New York.

Mott, N. F., Davis, E. A., and Street, R. A. (1975). *Phil. Mag.* **32**, 961.

Nakamura, Y., and Shimoji, M. (1969). *Trans. Faraday Soc.* **65**, 1509.

Nakamura, Y., and Shimoji, M. (1971). *Trans. Faraday Soc.* **67**, 1270.

Nakamura, Y., and Shimoji, M. (1973). *Proc. 2nd Int. Conf. Properties Liquid Met., 2nd, Tokyo* 567.

Nakamura, Y., Matsumura, K., and Shimoji, M. (1974). *J. Chem. Soc. Faraday Trans.* **70**, 273.

Nicotera, E., Corcia, M., deGiorgi, G., Villa, F., and Antonini, M. (1973). *J. Non-Cryst. Solids* **11**, 417.

Ninomiya, Y., Nakamura, Y., and Shimoji, M. (1973). *Phil. Mag.* **26**, 953.

Nyberg, D. W., and R. E. Burgess, (1962). *Can. J. Phys.* **40**, 1174.

Oberafo, A. A. (1975). *J. Phys. C.* **8**, 469.

Pauling, L. (1960). "The Nature of the Chemical Bond," 3rd ed. Cornell Univ. Press, Ithaca, New York.

Pearson, A. D. (1964). *J. Electrochem. Soc.* **111**, 783.

Perron, J. C. (1965). *C. R. Acad. Sci. Paris* **260**, 5760.

Perron, J. C. (1967). *Adv. Phys.* **16**, 657.

Perron, J. C. (1969). Thesis, D. Sci, Univ. of Paris, December 16.

Perron, J. C. (1970a). *Rev. Phys. Appl.* **5**, 611.

Perron, J. C. (1970b). *Phys. Lett.* **A32**, 169.

Perron, J. C. (1972). *J. Non-Cryst. Solids* **8–10**, 272.

Petit, R. B., and Camp, W. J. (1975). *Phys. Rev. Lett.* **35**, 182.

Phillips, J. C. (1973). "Bonds and Bands in Semiconductors." Academic Press, New York.

Popp, K., Tschirner, H-U., and Wobst, M. (1974). *Phil. Mag.* **30**, 685.

Ratti, V. K., and Bhatia, A. B. (1975). *J. Phys. F* **5**, 893.

Regel, A. R. (1948). *Zh. Tekhn. Fiz.* **18**, 1511.

Regel, A. R. *et al.* (1970). *Proc. Int. Conf. Phys. Semicond., 10th, Cambridge Massachusetts* 773.

Regel, A. R., Smirnov, I. A., and Shadrichev, E. V. (1971). *Phys. Status Solidi* **5**, 13.

Regel, A. R., Andreev, A. A., Kazandzhan, B. I., and Mamadaliev, (1972). *Sov. Phys.–Solid State* **13**, 2261.

Reitz, J. R. (1957). *Phys. Rev.* **105**, 1233.

Roll, A., and Motz, H. (1957). *Z. Metallk.* **48**, 272.

Roth, L. M. (1975). *Proc. Liquid Met. Conf., Mexico January 7–11, 1975.*

Ruska, J. (1973). *Proc. Int. Conf. Amorphous Liquid Semicond., 5th, Garmisch-Partenkirchen* 779.

Schaich, W., and Ashcroft, N. W. (1970). *Phys. Lett.* **A31**, 174.

Schmutzler, R. W., and Hensel, F. (1972). *Ber. Bunsenges.* **76**, 531.

Schmutzler, R. W., Fischer, R., Hoshino, H., and Hensel, F. (1976a). *Phys. Lett.* **A55**, 67.

Schmutzler, R. W., Hoshino, H., Fischer, R., and Hensel, F. (1976b). *Ber. Bunsenges.* **80**, 107.

Sen, P. N., and Cohen, M. H. (1972). *J. Non-Cryst. Solids* **8–10**, 147.

Seymour, E. F. W., and Brown, D. (1973). *Proc. Int. Conf. Properties Liquid Met. 2nd, Tokyo* 399.

Shadrichev, E. V., and Smirnov, I. A. (1971). *Sov. Phys.–Solid State* **12**, 2402.

Soven, P. (1967). *Phys. Rev.* **156**, 809.

Soven, P. (1969). *Phys. Rev.* **178**, 1136.

Steinleitner, G., and Freyland, W. (1975). *Phys. Lett.* **A55**, 163.

Steinleitner, G., Freyland, W., and Hensel, F. (1975). *Ber. Bunsenges.* **79**, 1186.

Stoneburner, D. F., Yang, L., and Derge, G. (1959). *Trans. AIME* **215**, 879.

Street, R. A., and Mott, N. F. (1975). *Phys. Rev. Lett.* **35**, 1293.

Styles, G. A. (1967). *Adv. Phys.* **16**, 275.

Takeuchi, S., and Endo, H. (1962). *Trans. J. Inst. Met.* **3**, 30.

Tauc, J. (ed.) (1974). "Amorphous and Liquid Semiconductors." Plenum Press, New York.

Tauc, J., and Abraham, A. (1968). *Helv. Phys. Acta* **41**, 1224.

Tauc, J., and Menth, A. (1972). *J. Non-Cryst. Solids* **8–10**, 569.

Taylor, P. C., Bishop, S. G., and Mitchell, D. L. (1971). *Phys. Rev. Lett.* **27**, 414.

Terpilowski, J., and Zaleska, E. (1963). *Rocz. Chem.* **37**, 193.

Terpilowski, J., and Zaleska, E. (1965). *Rocz. Chem.* **39**, 527.

Thorpe, M. F., and Weaire, D. (1973). *Proc. Int. Conf. Amorphous Liquid Semicond., 5th, Garmisch-Partenkirchen* 917.

Thouless, D. J., (1972). *J. Non-Cryst. Solids* **8–10**, 461.

Tolman, R. C. (1938). "The Principles of Statistical Mechanics." Oxford Univ. Press, London and New York.

Tosi, M. P. (1964). *Solid State Phys.* **16**, 1.

Tourand, G. (1975). *Phys. Lett.* **A54**, 209.

Tourand, G., and Breuil, M. (1970). *C. R. Acad. Sci. Paris* **B270** 109.

Tourand, G., Cabane, B., and Breuil, M. (1972). *J. Non-Cryst. Solids* **8–10**, 676.

Treusch, J., and Sandrock, R. (1966). *Phys. Status Solidi* **16**, 487.

Turner, R. (1973). *J. Phys. F* **3**, L57.

Urbain, G., and Übelacker, E. (1967). *Adv. Phys.* **16**, 429.

Valient, J. C., and Faber, T. E. (1974). *Phil. Mag.* **29**, 571.

Van Vleck, J. H. (1932). "The Theory of Electric and Magnetic Susceptibilities." Oxford Univ. Press, London and New York.

Vechko, I. N., Ditman, A. V., and Regel, A. R. (1968). *Sov. Phys.–Solid State* **9**, 2902.

Velicky, B., Kirkpatrick, S., and Ehrenreich, H. (1968). *Phys. Rev.* **175**, 747.

Vukalovich, M. P., Kazandzhan, B. I., and Chernyshev, R. N. (1967). Thermal Engineering, June. 95; (1967). *Teploenergetika* **14**, 70.

Warren, W. W. Jr. (1971). *Phys. Rev. B* **3**, 3708.

Warren, W. W. Jr. (1972a). *Phys. Rev. B* **6**, 2522.

Warren, W. W. Jr. (1972b). *J. Non-Cryst. Solids* **8–10**, 241.

Warren, W. W. Jr. (1973). *Proc. Int. Conf. Properties Liquid Metals, Tokyo 2nd* 395.

Warren, W. W. Jr., and Brennert, G. F. (1973). *Proc. Int. Conf. Amorphous Liquid Semicond., 5th, Garmisch-Partenkirchen* 1047.

Warren, W. W. Jr., and Clark, W. G. (1969). *Phys. Rev.* **177**, 600.

Weaire, D., and Thorpe, M. F. (1971). *Phys. Rev. B* **4**, 2508.

Wells, A. F. (1958). *Solid State Phys.* **7**, 426.

Wells, A. F. (1962). "Structural Inorganic Chemistry," 3rd ed. Oxford Univ. Press (Clarendon), London and New York.

White, R. M. (1974). *Phys. Rev. B* **10**, 3426.

Wilson, A. H. (1954). "The Theory of Metals," 2nd ed. Cambridge Univ. Press, London and New York.

Wilson, J. R. (1965). *Metall. Rev.* **10**, 381.

Yonezawa, F., Watabe, M., and Nakamura, M. (1974). *Phys. Rev. B* **10**, 2322.

Yosim, S. J., Grantham, L. F., Luchsinger, E. B., and Wike, R. (1963). *Rev. Sci. Instrum.* **34**, 994.

Zhuze, V. P., and Shelykh, A. I. (1965). *Sov. Phys.-Solid State* **7**, 942.

Ziman, J. M. (1961). *Phil. Mag.* **6**, 1013.

Ziman, J. M. (1967). *Adv. Phys.* **16**, 551.

INDEX